Maternities

Routledge International Studies of Women and Place

SERIES EDITORS

Janet Henshall Momsen and Janice Monk
University of California, Davis and University of Arizona, U.S.A.

1. Gender, Migration and Domestic Service
Edited by Janet Henshall Momsen

2. Gender Politics in the Asia-Pacific Region
Edited by Brenda S.A. Yeoh, Peggy Teo and Shirlena Huang

3. Geographies of Women's Health
Place, Diversity and Difference
Edited by Isabel Dyck, Nancy Davis Lewis and Sara McLafferty

4. Gender, Migration and the Dual Career Household
Irene Hardill

5. Female Sex Trafficking in Asia
The Resilience of Patriarchy in a Changing World
Vidyamali Samarasinghe

6. Gender and Landscape
Renegotiating the Moral Landscape
Edited by Lorraine Dowler, Josephine Carubia and Bonj Szczygiel

7. Maternities
Gender, Bodies and Space
Robyn Longhurst

Also available in this series:

Full Circles
Geographies of Women over the Life Course
Edited by Cindi Katz and Janice Monk

'Viva'
Women and Popular Protest in Latin America
Edited by Sarah A. Radcliffe and Sallie Westwood

Different Places, Different Voices
Gender and Development in Africa, Asia and Latin America
Edited by Janet Momsen and Vivian Kinnaird

Servicing the Middle Classes
Class, Gender and Waged Domestic Labour in Contemporary Britain
Nicky Gregson and Michelle Lowe

Women's Voices from the Rainforest
Janet Gabriel Townsend

Gender, Work and Space
Susan Hanson and Geraldine Pratt

Women and the Israeli Occupation
Edited by Tamar Mayer

Feminism / Postmodernism / Development
Edited by Marianne H. Marchand and Jane L. Parpart

Women of the European Union
The Politics of Work and Daily Life
Edited by Maria Dolors Garcia Ramon and Janice Monk

Who Will Mind the Baby?
Geographies of Childcare and Working Mothers
Edited by Kim England

Feminist Political Ecology
Global Issues and Local Experience
Edited by Dianne Rocheleau, Barbara Thomas-Slayter, and Esther Wangari

Women Divided
Gender, Religion and Politics in Northern Ireland
Rosemary Sales

Women's Lifeworlds
Women's Narratives on Shaping their Realities
Edited by Edith Sizoo

Gender, Planning and Human Rights
Edited by Tovi Fenster

Gender, Ethnicity and Place
Women and Identity in Guyana
Linda Peake and D. Alissa Trotz

Maternities

Gender, Bodies and Space

Robyn Longhurst

Routledge
Taylor & Francis Group
New York London

Routledge
Taylor & Francis Group
270 Madison Avenue
New York, NY 10016

Routledge
Taylor & Francis Group
2 Park Square
Milton Park, Abingdon
Oxon OX14 4RN

Routledge is an imprint of Taylor & Francis Group, an Informa business

Printed in the United States of America on acid-free paper
10 9 8 7 6 5 4 3 2 1

International Standard Book Number-13: 978-0-415-36046-3 (Hardcover)

Library of Congress Cataloging-in-Publication Data

Longhurst, Robyn, 1962-
Maternities : gender, bodies and space / Robyn Longhurst.
p. cm. -- (Routledge international studies of women and place)
Includes bibliographical references and index.
ISBN 978-0-415-36046-3 (hardback : alk. paper)
1. Motherhood. 2. Mothers--Social conditions. 3. Body, Human--Social aspects. 4. Pregnancy--Social aspects. 5. Human geography. 6. Feminist geography. I. Title.

HQ759.L576 2008
306.874'3--dc22 2007022779

Visit the Taylor & Francis Web site at
http://www.taylorandfrancis.com

and the Routledge Web site at
http://www.routledge.com

Contents

Illustrations

Preface

> Cells fuse, split, and proliferate; volumes grow, tissues stretch, and body fluids change rhythm, speeding up or slowing down. Within the body, growing a graft, indomitable, there is an other. And no one is present, within that simultaneously dual and alien space, to signify what is going on. "It happens, but I'm not there." "I cannot realize it, but it goes on." Motherhood's impossible syllogism. (Kristeva 1981: 237)

'Motherhood's impossible syllogism' continued to trouble me long after pregnancy. Even now, my sons aged 12 and 16, feel like they are part of me — utterly connected to me — yet totally separate from me, like alien beings or as Kristeva puts it 'alien space[s]'. At the risk of sounding clichéd, motherhood, for me, really has been a series of highs and lows, pain and pleasure — a schizophrenic journey of unimagined twists and turns.

I became a mother and an academic at the same time, beginning my PhD (on the subject of pregnancy) when I was pregnant with my first son. I have never been able (or ever really wanted to) separate these two major spheres of my life — motherhood and the academy. I have an ambivalent relationship with both. Perhaps quite appropriately this book then is the result of both my maternal and academic journey.

I first started thinking about maternities in 1992 when I enrolled in a doctoral thesis on pregnancy and public space. It has, therefore, been many years in conception and several years in writing. During a period of study leave in 2004 I decided to extend my thinking on pregnancy to consider related subjects such as birth, infant feeding, and the ways in which women and, indeed, men 'mother'. It was written in the midst of other projects and was often deflected by administration duties that I carried on a daily basis for three years as Chair of Department. The book is not necessarily intended to be read from beginning to end, from front to back. Rather, the reader can dip in and out. Some themes emerge in a number of the chapters. For example, the theme of 'queering maternal bodies' is shadowed in

Chapter 1, discussed briefly in Chapter 2 and elaborated on in Chapters 7 and 8. Other pieces do not fit together as readily.

From the comments above it is probably already apparent that the book is not a carefully crafted overview of motherhood or maternities, with one chapter leading directly to the next. There are already many such books that take this approach in a variety of disciplines. Rather this book aims to open up a series of windows on maternities. Each window provides a different view and yet all keep in sight the issues of gender, bodies and spatialities. One of the delights but also frustrations of working in the area of maternities is that it is a vast interdisciplinary area. I certainly don't profess to be an expert across all these fields. This book, therefore, won't offer readers everything they want to know about mothers, mothering and maternity. Its aims are more modest. It simply offers some windows, some views, maybe even just some glimpses, of this complex and rich area of study. Nevertheless my hope is that readers will find these glimpses intriguing and they will stimulate more interest in this important topic.

Acknowledgments

Academic work is a collective enterprise. Conversations with friends and colleagues over lunch, my partner shouldering more than his fair share of the domestic load, and my mother willingly and lovingly taking care of the children at every opportunity have all helped tremendously in enabling me to complete this book. My debts to individuals are numerous, too numerous in fact to list in full, but there are a number which in memory stand out.

I owe special thanks to my friend and colleague Lynda Johnston. Sharing academic life with Lynda is a pleasurable and an intellectually rewarding experience. Her contributions over many years and her more recent contributions such as reading a draft of this book have been much appreciated. John Campbell has also been a valued friend and colleague for more than a decade. Although we work in different fields within the discipline of geography, John's political and social commentary on a variety of issues often helps me think through issues. Kathryn Besio during her time at Waikato from 2001–2004 was also a valued friend and colleague. Co-teaching, co-researching, and chatting with KB over lunch always left me with something useful to ponder. Richard Bedford also deserves special thanks. Dick has mentored me in all aspects of academic life and I am most grateful.

I also want to thank my colleagues in the Department of Geography, Tourism and Environmental Planning at the University of Waikato. They are a fascinating bunch of people with whom I share many convivial morning and afternoon tea breaks. During the time I wrote this book I was serving a term as Chairperson of Department plus carrying my usual teaching load. This did not make writing easy but many of my colleagues assisted wherever and whenever possible.

Amanda Banks, a Masters graduate and research assistant in the Department of Geography, Tourism and Environmental Planning, was the best student and research assistant anyone could ever wish for. Amanda never tired of locating excellent resource material for this project. Without her efforts the book would have taken much longer to write and the researching and writing would not have been as pleasurable. Betty-Ann Kamp, a cartographer in the Department, assisted with developing plates. I am most appreciative.

There are a number of other people who have also assisted, albeit at different times, and in different ways. I am grateful to Margaret Begg, Elaine Bliss, Liz Bondi, Lex Chalmers, Cathy Coleborne, Tovi Fenster, Brenda Hall, Anne-Marie d'Hautesserre, Sara Kindon, Russell Kirkpatrick, Audrey Kobayashi, Kyro Selket, Yvonne Underhill-Sem, Ottilie Stolte, Brenda Yeoh, and Dan Zirker. All of these people have assisted me in one way or another and I offer them my sincere thanks.

I am also much indebted to the many participants who took part in this study. Their stories provided the basis for some of the chapters. It was a privilege to be privy to their lives as mothers. I am especially grateful to Angeline Greensill who shared with me her 'birthing history'. Over the years Angeline's insights on mothering and on *taha* Māori, the environment, and politics have proven invaluable to me and I am sure to many others.

Thanks also to the series editors Janice Monk and Janet Momsen who have offered support throughout this project. Their understanding and belief that the book was worthwhile and would be completed (in the end) was most appreciated. Thanks also to Terry Clague and Benjamin Holtzman, Routledge Research and Commissioning Editors.

Last but certainly not least, a huge thank you must go to my nearest and dearest — to my partner David for all the meals cooked while I worked on this book, to my children Benjamin and Jerome who each day teach me something new, and most importantly to my mother Colleen, to whom this book is dedicated. To all who have nurtured me and continue to nurture me in a myriad of ways, I offer my deep and lasting appreciation.

A similar version of Chapter 3 was published in *Gender, Place and Culture: A Journal of Feminist Geography,* Vol. 12, No. 4, pp. 433–446, December 2005. I should like to thank the editors and publisher for permission to reproduce it. Chapter 5 was published in *ACME,* Vol. 5, No. 2, pp. 209–229, 2006. I should also like to thank the editors and publisher for permission to reproduce it. A shortened version of Chapter 7 was presented at the International Geographical Union (IGU) Commission on Gender and Geography Pre-Congress Symposium 'Shifting Boundaries: Gender, Bodies and Spaces', hosted by the Department of Geography, Tourism and Environmental Planning at the Waikato Museum of Art and History, 28–30 June 2006. I am grateful to all those who offered invaluable feedback on the presentation.

The author and publishers also wish to thank Rod Emmerson and the *New Zealand Herald* for allowing us to reproduce the 'Breast-fed as a pup cartoon', the *New Zealand Herald* for use of the photograph of 'Kura Tumanako and her breastfed pup Honey Boy', and Allan Hawkey and *Waikato Times* for use of the 'Large litter caroon'. Finally, thanks are due to the Hamilton City Council which offers an excellent Online Image Library of high quality downloadable images that are freely available for private and professional use.

Every effort has been made to trace copyright holders but if any have been inadvertently overlooked, the publisher will be pleased to make the necessary arrangements at the first opportunity.

1 A Series of Windows

This book provides a series of windows on maternities highlighting issues of gender, bodies and space. It demonstrates the multiple ways in which mothering and maternal bodies influence and are influenced by social and cultural processes and the politics of space. Over the past two decades geographers have shown a growing interest in bodies, including gendered bodies, as an important co-ordinate of subjectivity and as a way of understanding further relationships between people, place and space. There have been numerous sessions on bodies at conferences such as the Association of American Geographers. Copious articles on bodies have been published in geographical journals such as *Gender, Place and Culture*. Edited collections such as Elizabeth Teather's (1999) *Embodied Geographies* and authored books such as Steve Pile's (1996) *The Body and the City* can now be found on library shelves. In short, there has been a burgeoning of literature on bodies and space.

Attention has been focused on the ways in which embodied subjectivity and spatiality are intimately entwined (Probyn 2003). Bodies are always located (Longhurst 2001a; Moss and Dyck 2002; Nast and Pile 1998) and interpellated by a range of discourses and ideological systems. People inhabit different subjectivities (sometimes contradictory) in different spaces. Bodies both produce space and are produced by space. Bodies and space rub against, fold, and sometimes disintegrate (see Davidson 2003 on agoraphobia) into each other.

Much of this literature on bodies and space necessarily involves a discussion of gender because bodies are not simply human bodies, they are gendered bodies (Gatens 1991: 82; also see Davis 1997). Over the past decade feminist geographers have embarked upon much research on gender relations and gendered bodies (for example see Duncan 1996; Johnson 1989; Johnston 1997; McDowell and Court 1994; Moss and Dyck 2002; Nelson 1999; Pratt with the Philippine Women's Center 1998). This research has challenged processes that lead to inequalities between men and women (Massey 1994; Rose G 1993). But even more than this, feminist geographers have reflected 'on what the categories "women" and "men" mean, and on the concept of gender, in the context of social identities and social relations more generally' (Bondi and Davidson 2005: 15).

Despite the attention paid to gender, to date, geographers have published little on what surely must be one of, if not *the*, most important of all bodies — bodies that conceive, give birth, and nurture other bodies — that is, maternal bodies (but there are exceptions such as Madge, Noxolo and Raghuram 2004; Madge and O'Connor 2005; Mahon-Daly and Andrews 2002). This absence in the geographical literature is all the more surprising given that over the past decade work on maternities by feminists, psychologists, anthropologists and sociologists has expanded dramatically.

> Over the past decade, the literature on maternity and motherhood has expanded dramatically. ... feminist scholarship has turned from the initial problem of the role of maternity in defining femaleness to interrogating its place in the social relations between men and women and among women, the meanings with which maternity as experience and status are imbued, and the tensions and accommodations of reproduction and production institutionally and personally in different places and under different systems of production. (Manderson 1998: 26)

Lenore Manderson (1998) makes an important point. Maternities and motherhood are produced differently in different places. Geographer Sarah Holloway (1999: 91) argues 'Far from being a simply natural experience, motherhood is a complex social phenomenon: it varies over time and space, and is intimately bound up with normative ideas about femininity.'

This inextricability of bodies, gender, place and culture can be demonstrated through the example of Māori women in New Zealand. Like all bodies, Māori women's bodies cannot be understood outside of the cultural practices and spaces within which they are constructed, and in turn construct. For example, Wikitoria August (2005) explains that when Māori women are *hapū* (pregnant) they face restrictions from cultural spaces such as cemeteries and sites of food gathering. August (2005) argues that these restrictions, however, cannot simply be read through a Western framework as exclusion or marginalisation. Māori women, rather than feeling restricted from particular spaces, often feel that they are showing respect for particular spaces, and a respect for their pregnant bodies and for *tikanga* (customs).

Understanding maternities through different cultural lenses does not deny the fact that there is a 'real' material or physical body — a body in which sperm travels up a Fallopian tube, an egg is fertilized, a fetus grows and eventually is born. Rather, it suggests that there is not simply a 'real' material body on the one hand and its various cultural representations on the other. In other words, representations, understandings and cultural inscriptions quite literally constitute maternal bodies and help to produce them as such. Maternal bodies are socially, sexually, ethnically, class specific bodies that are mutable in terms of their cultural production. It is important to extend the notion of physicality that dominates biological and medical sciences to illustrate that bodies are materialities that are not containable in physical terms alone. This is not to suggest that biological, medical, and popularist accounts or analyses of maternities are 'wrong' but rather that the guiding assumptions and prevailing methods used by these writers and researchers have tangible effects on maternal bodies.

It is timely, therefore, for feminist, social and cultural geographers to contribute more to these debates about maternal bodies, spaces and places. Examining maternities through social and spatial concepts common to many geographers (such as radical revisions of nature, cultural (re)production and relations, shifting moral boundaries, and deconstructions of binaries such as private/public and in the closet/out of the closet) is potentially rich. Such an approach offers opportunities to rethink maternal bodies as inseparable from the spaces and places of their (re)production.

In short then this book explores the diversity and complexity of embodied experiences of maternity and illustrates that maternal bodies are constructed differently through different social and cultural networks and through different spaces and places. My hope is that it will add to the geography literature by addressing previously ignored bodies and spaces and that it will add to the maternities literature by offering a spatial perspective. In other words, maternal bodies need to be examined further through a spatial lens. Geographers have a great deal to offer those in other disciplines by adding space as a critical component in the construction of specific embodied performances (Butler 1990).

MATERNAL BODIES

This research is positioned simultaneously both in social feminist geographical literatures on bodies and in a broader interdisciplinary literature on mothers, mothering, and maternal bodies (a literature that often combines health, disability, sociology, anthropology, psychology, psychoanalysis, and cultural studies). Like Margaret Jolly (1998), a historical anthropologist who has written extensively on gender in the Pacific, I tend to use the term 'maternities' rather than 'mothering.' Jolly (1998: 1) explains she made this decision in the edited collection *Maternities and Modernities* (Ram and Jolly 1998) to 'highlight the corporeal processes of being pregnant, giving birth and nurturing'. Jolly (1998: 2) is quick to point out, however, that these 'corporeal processes' or 'seemingly natural processes' of 'swelling, bearing and suckling, the flows of blood, semen and milk are constituted and fixed not just by the force of cultural conception but by coagulations of power.'

It is important to remember, as Sara Ruddick (1989: 48) notes, 'Pregnancy, birth, and lactation are different in kind from other maternal work and, measured by the life of one child, are brief episodes in years of mothering.' Other kinds of mothering work such as looking after a toddler, helping children with homework, guiding a teenager through a complex ethical dilemma are perhaps more characteristic of the demands on a mother than carrying a fetus, or feeding an infant. Nevertheless, pregnancy, birth and lactation are often experienced as profound and life-changing events that prompt careful consideration of bodies and space (see Teather 1999 on rites of passage) and therefore this book tends to focus on some of the initial journeys into motherhood.

'New' mothers or 'becoming' mothers — maternal bodies engaged in the production of new selves — are useful to examine because they are often the most acutely aware of their changed corporeality and social status as they begin to confront what it means to be a 'mother'. It is acknowledged that elderly women still 'mother' their adult children, and 'grandmother' their grandchildren but this book focuses mainly on the processes involved in taking up maternal subject positions. It makes an argument for understanding maternal bodies as historically and geographically constituted, for understanding the social and spatial processes that help make mothers.

Hopefully it has become apparent by now that maternal bodies are not simply sets of biomedical facts which are gradually being uncovered by science (see Morgan and Scott 1993: 7–10). What it means to be a mother shifts across time and space. The signifier 'maternity' — which has been derived from the fifteenth century Medieval Latin *māternālis*, from Latin *māternus*, from *māter* mother — is not fixed (*Collins English Dictionary* 1979: 909). The word maternity carries with it several different meanings that are used in different contexts at different times. First, it is sometimes used to refer to the state of being pregnant; the period from conception to birth when a woman carries a developing fetus in her uterus. Second, it is used to refer to the kinship relation between an offspring and the mother. Third, the term is used to refer to the quality of having or showing the tenderness and warmth and affection of or befitting a 'proper' or 'good' mother.

The book draws on all these meanings. It aims to examine the embodied experiences of mothers but also the ways in which maternities are constituted in discourses. Maternal bodies, contrary to popular belief, are not entirely 'natural', rather they are an interface between nature and culture, biology and the social, materiality and discourse. They are always located in time and space, that is, they have histories and geographies. In other writing (Longhurst 2001a) I have problematized pregnant bodies arguing that 'It is not possible simply to add pregnancy to a body, nor is it possible simply to add a body to something called pregnancy' (Longhurst 2001a: 3). Pregnant bodies trouble binary thinking. They undergo a bodily process that transgresses the boundary between inside and outside, self and other, subject and object, and often female and male (when women carry boys). Likewise, it is not possible simply to add something called maternity to a body nor is it possible simply to add a body to something called maternity. Discourses and materialities of maternity vary greatly over space but also time.

Older mothers aged in their 60s and beyond, for example, often talk about how different life was for them than for their daughters. In New Zealand in the 1950s and 1960s the dictum 'spare the rod, spoil the child' was common. Mothers often hit their children believing that such discipline was required to ensure their children would not grow up 'spoilt'. In 2005 a Member of Parliament in New Zealand put forward a bill to repeal Section 59 of the New Zealand Crimes Act that allows a parent to defend an assault charge by arguing that he or she was using 'reasonable force' in physically punishing a child (Tunnah 2005). The Section of the Act has been heavily criticized in high-profile cases of child abuse with claims being made that allowing a parent to use such a defense for beating a child is 'barbaric'. Others think that the law requires no repealing (see Christians vow to break smacking law, 9 June 2006, available <http://www.nzherald.co.nz>, accessed 29 August 2006).

What this example illustrates is that the discourses that surround hitting or smacking children are changing. Smacking is now less likely to be seen as a maternal act that is in the child's best interest. Maternal performances of disciplining children's bodies change over time. 'Positive parenting techniques' such as 'time out' are now seen as preferable to smacking.

There is no stable foundation, no essence, to being a mother or to the practices surrounding motherhood. Claire Madge and Henrietta O'Connor (2005: 94) explain:

> There is no preconstituted 'body' on to which motherhood is inscribed; what it means to be a mother is constantly produced and reproduced through varying and competing discourses and practices from a variety of different people and places. The construction of mother as a category is not a pre-given, coherent and stable subject position.

If the category of mother is not pre-given, this raises the question can anyone become a mother? Can a woman who has not given birth become mother? Can a man become a mother? Society deems some women to be 'inappropriate' mothers — too selfish, too old, too young, too queer, and/or have too many children already. Not all women want to have children or to be mothers. Some men, however, embrace the gendered expectations of mothering and delight in nurturing, fostering and taking care of children (see Doucet 2006 on men mothering). As Stuart Aitken (2000: 581) suggests 'Our enduring myths of social reproduction do not seem to support forms of masculinity that encompass a sense of self that is nurturing and domestically orientated' but that is not to say that men cannot and do not mother.

DECONSTRUCTING MATERNITIES AND PATERNITIES

In 2005 I asked a class of geography graduate students what comes to mind when you think about mothers and mothering? Most answered that mothering speaks to a *woman's* relationships with her child or children usually involving unconditional love, commitment, vast responsibilities, sacrifice, nurturing, comforting, protecting, disciplining, educating, feeding and toileting. A few students also noted that sometimes mothers treat their children badly. They abuse, hurt, suffocate, maybe even kill or abandon their children. Continuing this conversation, I asked students, given that men can love, nurture and comfort children, is it possible for men to become mothers? Their responses were overwhelmingly 'no'. One student asked 'wouldn't men who mother — nurture and comfort — be fathers?' This is an excellent question that gets to the heart of attempts to understand and problematise maternities, gender, bodies and space.

Aitken also ponders this question and offers the following insight:

> There is a strange immiscibility between fathering and child care that I am not quite sure how to reconcile. Can it be that part of the problem relates to men's inability to bear children and provide them with 'natural' sustenance as infants? To suggest such is to return to a form of naturalism that is groundless in light of contemporary feminist and poststructuralist critiques ... our contemporary critiques of empiricism suggest that no fact — such as ability to bear and nurse children — entails or excludes a moral right or commitment. As a father, then, I am comforted and liberated by the idea that gender is a socially imposed division of the sexes, and that there is no 'natural opposition' between men and women. But current fatherhood ideology and social science research continues to define a father's relationship with his children as co-parenting that is interdependent with, and sometimes in opposition to, mothering. (Aitken 2000: 581–582)

I agree with Aitken that the gender performances surrounding care-giving tend to be constructed differently for men and women. This is not surprising given that the terms 'maternity' and 'paternity' reverberate differently (despite the fact that mothers and fathers are both parents). Paternity is described in the *Collins English Dictionary* (1979: 1075) as being derived from the Latin *pater*, meaning father. Whereas mothers have long been associated with traits such as love, nurturing and affection, fathers have long been associated with traits such as origin, responsibility and respect. Some common phrases that include the term 'father' are Church Father, Father Time, and Father Christmas. Some common phrases that include the term 'mother' are mother country, mother tongue, Mother Goose, Mother Hubbard and, the more derogatory, mother fucker. While maternity and paternity both fall under the umbrella of 'parenthood', each has a different resonance that attaches to what are seen to be 'appropriately' sexed and gendered bodily performances. Supposedly mothers (women) are maternal while fathers (men) are paternal. To construct a binary division between these terms, however, is unhelpful in that it shuts down a variety of rich and complex subjectivities inhabited by women and men.

To return for a moment to the conversation with the graduate students, the trouble they had connecting motherhood and men reflects a widespread belief in the West and beyond that mothering is natural and connected to essential femininity. Sarah Holloway (1999: 91) says 'Motherhood is often regarded as a natural phenomenon: it is looked upon as normal for a woman to want to bear and raise children'. Terms such as women/men, maternity/paternity, and mothers/fathers are locked in a binary relationship. Without wanting to rehash already well rehearsed arguments, post-structuralist feminists such as Elizabeth Grosz (1989: xvi) have explained that when a continuous spectrum is divided into discrete self-contained elements which exist in opposition to each other, this is known as a dualism, dichotomy or binary. 'Dualism is the belief that there are two mutually exclusive types of "thing" . . . that compose the universe in general and subjectivity in particular' (Grosz 1994: vii), for example, white/black, good/bad, rich/poor, light/darkness, mental/physical, mind/body. Pamela Moss and Isabel Dyck (2002: 13) add to this:

> What is so insidious about dualisms is that they permeate all realms of thought and pose, support, and reproduce a polarized understanding of what it is to exist, to know, and to act. By providing only 'either/or' patterns for understanding bodies, dualisms systematically relegate bodily knowledge, as well as the bodies themselves, into multiple sets of binarisms.

This book aims to present a way of understanding bodies as sites where *both* mind and body, rationality and irrationality, masculinity and femininity, and most importantly, paternity and maternity, are simultaneously located. It is important to dislodge the simplistic alignment of women with motherhood and maternity, and men with fatherhood and paternity because these categories are ambiguous and encompass their own opposites. Hopefully, this will become apparent in the book.

In order to trouble dualisms such as man/woman and father/mother I draw on a long tradition. Numerous theorists have argued for the radical and productive possibilities of transforming the 'or' in binary pairs into 'and'. For example, Jacques Derrida (1981) in 'Plato's Pharmacy' uses the Greek term *pharmakon* (meaning medicine as a means for healing and/or poison as a means of killing) to discuss ambiguities — categories can take in their own opposites (see Moss and Dyck 2002: 13). Both Moira Gatens (1988) and Grosz (1994) argue that seventeenth century philosopher Benedictus Spinoza may provide a way to bypass the dualisms which dominate traditional philosophy. Spinoza understood the mind and body not as separate but rather the mind as an idea of the body. Spinoza worked with a notion of difference that was non-oppositional. Homi Bhabha argues for the notion of 'Third Space'[1] (see Bhabha 1990, 1994). Bhabha (1994: 39 cited in Longhurst 2001b: 124) explains 'by exploring this Third Space, we may elude the politics of polarity and emerge as the others of our selves'. Julia Kristeva draws on what Plato in the *Timaeus* calls the *chora* to explain the semiotic. She says the chora is 'receptacle, unnameable, improbable, hybrid, anterior to naming, to the one, to the father and consequently maternally connoted' (Kristeva 1980: 133). The *chora* is the site of the undifferentiated bodily space the mother and child share. 'A site for the production of the matrix/womb and matter, the *chora* is the unnameable, unspeakable corporeality of the inextricably tangled mother/child dyad which makes the semiotic possible' (Wright 1992: 195).

But perhaps what needs to be inextricably tangled is not only the 'unspeakable corporeality' of mother and child but also of mother, father and child or maybe mother, mother and child or maybe father, father and child. Or why not father, friend and child or an infinite array of other (including non-human — see Chapter 7 on interspecies breastfeeding) possibilities that exist around maternity and paternity? Entangling and reconfiguring maternities as in-between and multiple embodied discursive spaces offers a productive way of thinking about the complexities of maternal relationships and de-essentialising 'the mother's body'.

To think through this idea a little further it is useful to consider the following questions. Is the 'wet nurse' who provides breast milk for a child that the birth mother cannot or does not wish to feed, a maternal body? Is the man who cares for, nurtures, and loves his children, a maternal body? Is the woman who suckles a young pig or a puppy, a maternal body? (see Giles 2004 on 'queer breastfeeding' and Chapter 7 this volume). Is the woman who has no (biological) children of her own but tends to the emotional and physical needs of other children, a maternal body? Is the woman who neglects, mistreats or kills her child, a maternal body? Is the older sibling who takes great care of his or her younger sibling(s) (sometimes in the absence of a biological mother), a maternal body? Is the man who experiences physical aspects of his partner's pregnancy and/or birth (known as the Couvade syndrome), a maternal body? Is the woman who exhibits signs of pregnancy such as tiredness, morning sickness, tenderness of the breasts, and perhaps even a swelling of the abdomen and yet is medically diagnosed as *not* pregnant (referred to as phantom pregnancies or pseudocyesis), a maternal body? Is the man who wears a 'sympathy belly' when his female partner or wife is pregnant, a maternal body? The list of questions goes on.

The point being made here is if there is no pre-constituted, essential maternal body and instead mothering can be attached to numerous different bodies, in different times and spaces, then the answers to these questions are necessarily complex. The questions highlight that while not denying the 'real' fleshy materiality of bodies it is important to understand maternities as socially, historically, and spatially constituted.

In this book maternal bodies are examined using a version of feminist poststructuralist theory that holds on firmly to the notion of materiality and discourse. In Judith Butler's (1990: 33) influential book *Gender Trouble* she argues that gender is 'the repeated stylization of the body, a set of repeated acts within a highly rigid frame that congeal over time to produce the appearance of substance, of a natural sort of being.' Mothering is a gendered performance, a set of repeated citational and bodily acts that produce the appearance of the 'natural' and the 'true'. Repeated performances of expected maternal behaviours establish regulatory practices for mothers. Butler displaces the heteronormative alignment of gender and sex by arguing that gender is 'performative'. It is also possible to displace the alignment of maternity (a set of gendered acts) and a (sexed) body. Butler argues that the gendered body 'has no ontological status apart from the various acts which constitute its reality ... if that reality is fabricated as an interior essence, that very interiority is an effect and function of a decidedly public and social discourse' (Butler 1990: 136). If maternal bodies have no essence then anyone can engage in mothering and become a mother.

A NOTE ON THEORIES AND METHODS

The theoretical and methodological approach adopted for the purposes of this book grew out of many years of conversations, interviews and focus groups with mothers and becoming mothers, in particular, pregnant women in Hamilton, New Zealand. The empirical data generated by these conversations, interviews and focus groups is not drawn on in every chapter but it has helped me develop a kind of poststructuralist feminist analytical framework that holds on fiercely to lived experiences, a framework that is similar to Moss and Dyck's (2002) notion of a 'radical body politics'. Bodies cannot be reduced to a set of discursive formations. They are also material entities that hurt when pinched and bleed when cut. As Jan Draper (2003: 746) notes 'Emphasis on "body theory" has, in a paradoxical way, left bodies out' (Davis 1997; Leder 1990; Longhurst 2001b). In relation to the discipline of geography, elsewhere I have argued

> The messy surfaces/depths of bodies, their insecure boundaries, the fluids that seep and leak from them, that which they engulf, the insides and outsides that sometimes collapse into each other, remain invisible in the geographical canon ... The reason this is significant is that the messiness of bodies is often conceptualised as feminised and as such is Othered. Bracketing out questions about the boundaries of body/space relationships functions as an attempt to position geographical knowledge as that which can be separated out from corporeality, the

> corporeality of its subjects and its producers. Ignoring the messy body is not a harmless omission, rather it contains a political imperative that helps keep masculinism intact. (Longhurst 2001b: 23)

Over the past 15 years I have collected numerous materials on pregnancy and on motherhood more generally. I have a small library of academic materials and popular materials such as newspaper and magazine articles, information downloaded from Internet sites, photos, films, videos, and documentaries on mothering. These materials have been collected in New Zealand but also in range of other places including the United States, the United Kingdom, Europe, Asia and the Pacific. The mainly qualitative data, and processes used to collect them, varies from chapter to chapter. Details about the research methods used in Chapter 2 and Chapter 3 are presented in the Appendix. Details about the research methods used in the other chapters are presented near the beginning of each chapter.

LITERATURE REVIEW

The literature on maternities, maternal bodies and mothering is vast, too vast to review here. I have focused, therefore on that work that interests me most as a poststructuralist feminist geographer. This tends to be work that has been written by geographers, and others interested in issues such as space, place, spatialities, scale, territories, borders, margins and boundaries. Some of this material is cited and discussed in other chapters in the book but it's useful to synthesize it here in order to provide readers with a ready package of material on geographical work on maternal bodies. The review is structured loosely around a series of geographical scales from the body to the nation. Some of the research, however, cuts across a variety of geographical scales and therefore it is difficult to know where to include it. This research, to use Neil Smith's (1993) term, 'jumps scale'. As Gill Valentine (2001) notes there is not anything 'natural' or inevitable about geographical scales. Different scales simply provide different platforms for examining socio-spatial scales.

At the micro-scale, the space of the body itself, some interesting research has been carried out on pregnant bodies, which are usually assumed to be maternal. One of the most influential and often cited contributions on pregnant embodiment is Iris Marion Young's (1990) phenomenological exploration of the pregnant subject. Pregnant, birthing and lactating bodies all destabilize binaries between mother and infant, one and two, inside and outside, self and other. In relation to pregnancy Young (1990: 163) explains 'Pregnancy challenges the integration of my body experience by rendering fluid the boundary between what is within, myself, and what is outside, separate. I experience my insides as the space of another, yet my own body.' During pregnancy the borders of the body change rapidly

— they are not fixed — but rather are volatile. Women often talk about their pregnant stomachs rubbing against, and bumping into, things in confined spaces illustrating how bodies and spaces rub, fold and sometimes disintegrate into each other. They find their enlarged stomachs where they do not expect to because their material bodies have grown rapidly and it often takes some time before their body image catches up (see Grosz 1994: 79 on the ways in which zones surrounding bodies change).

Kristeva (1981: 31) notes 'Pregnancy seems to be experienced as the radical ordeal of the splitting of the subject; redoubling up of the body, separation and coexistence of the self and an other, of nature and consciousness, of physiology and speech.' But it is not only pregnant bodies that are volatile and cannot be taken for granted. Birthing and breastfeeding bodies also challenge notions of a stable, autonomous, self-contained body. These bodies do not function as solely discrete or individual bodies (see Bailey 1999). Draper (2003: 748) notes

> Some women enjoy this unique intimacy with their baby, others however, may never come to terms with this tension (Schmied and Lupton 2001a) and can feel devoured or invaded by the baby. (Lupton and Barclay 1997; Young 1990)

Lupton and Barclay (1997: 32 cited in Draper 2003: 748 — Draper's italics) explain this blurring, and even breaking (for example, in labour and lactation) of boundaries 'may be experienced as more confronting by men because it challenges specific ideals of masculinity. *These ideals tend to position the male body/self as far more separate and autonomous than the female body/self.*'

Still at the micro-scale of the body itself, and still focusing on the pregnant body as a site and symbol of maternity, Tyler (2001) explores the skin of the pregnant body and the unborn baby using the psychoanalytic concept of 'skin ego'. She argues that the photographic feature in *Vanity Fair* (August 1991) of the heavily pregnant Demi Moore dramatically changed public representations of pregnant women in the West through a rethinking of the cultural taboo on pregnant women. The maternal subject became able to simultaneously take up a new subject position, that of 'sexy' woman.

Moving spatial scales now to consider maternal bodies not as individuals but as social groups, Jaqui Gabb (2004) analyzes lesbian mothers. She examines passion and desire in lesbian mother–children love. There are many types of love which differ in character and expression. Gabb (2004: 399: italics in original) considers 'the *physicality* of the mother–child relationship, the corporeal elements of sensuality and passion that constitute the embodied experience of maternal–children intimacy' for lesbians. Gabb's insights on the mother–child relationship for her particular group of mothers are informed by the psychoanalytical work of Sigmund

Freud, Melanie Klein, Jean Lapanche, Teresa de Lauretis, Julia Kristeva and Nancy Chodorow.

Lucy Bailey (2001) examines a different group of mothers. She draws on data collected from 30 first-time pregnant middle-class women in Bristol, England, and explores the tensions and difference expressed by the women interviewed that arose after the birth of their child. She identifies three bodily aspects of women's gendered sense of self — sensuality, shape, and space. In relation to space, interestingly, Bailey found that in many cases, the pregnant women whom she interviewed were afforded *more* public space. Some of Bailey's respondents claimed that people stood back to let them go first on trains or buses. However, the increase in space accorded pregnant women is troubled by its dissolution in other spaces such as the workplace. Bailey (2001: 123) argues 'Whereas the women felt more valued and given more bodily space in the public arena, in their worlds of employment, the pregnancy could regender them as workers in undermining ways.'

Holloway (1999) also examines the mothering practices of 'white' mothers in the United Kingdom. She presents a case study that focuses on two suburbs in Sheffield, UK, to explore how different groups of mothers, in particular 'white' heterosexual mothers from working and middle-class backgrounds negotiate motherhood. Holloway argues that mothering practices are shaped within the context of 'local child care cultures' which enable and/or constrain mothers' choices about how to organize their everyday lives (also see Holloway 1998a, 1998b, 1998c, 1999).

Ekinsmyth, Elmhirst, Holloway, and Jarvis (2004) also consider cultures surrounding mothering, child care and workplaces. They each provide a reflexive, autobiographical account of the ways in which they combine academic work and parenting/mothering. Ekinsmyth *et al.* (2004: 95) are concerned with 'reflecting on the ways in which our feminisms have been transformed as we have been changed as women by the experience of motherhood.' The experience of motherhood not only prompts the authors to re-examine the politics of research but also makes them aware of the lack of academic scholarship on geographies of child rearing and domestic life. While highlighting the diversity of their individual experiences of mothering and academia, they collectively call for a 're-valuing, valuing or valorizing of the domestic, the home and the role of the mother' (Eckinsmyth *et al.* 2004: 104). They usefully draw attention to workplace practices within the academy.

While some researchers have addressed the spaces of homes, public transport and workplaces for mothers, others have addressed the spaces of birth. Scott Sharpe (1999) illustrates that the popular (Western) place of birth — the birthing unit — is discursively constructed through language. Maria Fannin (2003) also examines a birthing unit but in the United States. She argues that home and hospital have long been shifting and contested sites of childbirth in the United States. Fannin examines the transformation of hospital space into 'homelike' birthing rooms — a transformation that has come about as one of the effects of deregulation and downsizing of U.S. health care over the past few years. The boundaries between market, hospital and home have been blurred by neoliberal economic logics as hospitals have begun to market a particular kind of birth to middle-class 'consumers'. Sally Abel and Robin Kearns (1991) also focus on the place of birth and offer a geographical perspective on planned home birth in New Zealand. Abel and Kearns (1991) conducted a qualitative study of six women's options and experiences of home birth. Abel also drew on her experiences as a midwife. The women all lived in Auckland. The findings of the study are discussed under three themes: accessibility of home birth as an option (three of the participants were actively discouraged by their General Practitioner (GP) from having a home birth), reasons for choosing a home birth (including feeling in control and having continuity of care) and the experience of home (this material on the place of birth informs Chapter 6).

Still other researchers (especially in the United Kingdom) have paid attention to issues of breastfeeding and space. Patricia Mahon-Daly and Gavin Andrews (2002) discuss how breastfeeding creates liminality, an 'in-between-ness', in relation to time, space and the body. They examine how place impacts on breastfeeding behaviour and how liminality can sometimes be utilized by breastfeeding women. Mahon-Daly and Andrews (2002) also discuss the 'leakiness' of breast milk and the ways in which breastfeeding transgresses bodily boundaries. This is one possible reason why breastfeeding rates in the industrialized world are low. Rachel Pain, Cathy Bailey and Graham Mowl (2001) carried out a qualitative pilot study of the infant feeding decisions of first-time parents in Newcastle upon Tyne, England. Pain *et al.* (2001) were keen to find out why Britain has comparably higher rates of formula feeding. Findings illustrate that gendered ideologies of parenting and the ways in which these different ideologies are constituted in and through different spaces such as the home, workplace and public space influence infant feeding decisions. Cultural shifts in attitudes towards breastfeeding are required to increase breastfeeding rates rather than simply reiterating the rationale that 'breast is best' (this material on the breastfeeding and space informs Chapter 7).

Turning attention now from the scale of local communities to the scale of the nation, Lily Phua (1998) examines power relations and the politics of reproduction in Singapore. Pregnancy in Singapore, she argues, is infused with state, medical and cultural ideologies. Women, at times, resist these ideologies in an attempt to maintain control over their own bodies. Phua contrasts the regulation of pregnant bodies under Western biomedicine and traditional Chinese medicine (also see Phua and Yeoh 2002 on pregnancy in Singapore).

Yvonne Underhill-Sem (2000) examines maternal bodies at the level of a village — Wanigela — but situates this within a broader discussion of cultural practices in Papua New Guinea (PNG). Underhill-Sem challenges traditional notions of population geography by drawing attention to the specificities of maternities in Wanigela, Oro Province, PNG. She draws on Jolly's (1998) concept of maternities which takes into account women who do not necessarily bear children and includes women who are involved in mothering activities or women who become mothers through other means.

Madge and O'Connor (2005) research maternities using yet another spatial scale, a scale that transcends national boundaries. They examine cyberspace. Madge and O'Connor argue that Judith Butler's (1993) highly influential work on performance and performativity has prompted geographers (e.g. Bell *et al.* 1994; Gregson and Rose 2000; Nash 2000; Nelson 1999) to think carefully about the potential of these concepts for understanding further issues of spatiality, especially in relation to gender and sexuality. Madge and O'Connor (2005) examine how new mothers (e-mums) go online to try out new subject positions.

> These performative aspects of mothering identity are complex and uncertain, but it is through the specific performances of mothering identity that cyberspace is brought into being: cyberspace is actualized through performance. (Madge and O'Connor 2005: 85)

Madge and O'Connor's (2005 and 2006) work has been important in informing Chapter 9.

This review barely scratches the surface of the rich corpus of work that has been carried out to date on maternal bodies and maternities. For example, no mention has been made of some the feminist 'greats' — the foundation texts on the subject — books such as Shulamith Firestone's *The Dialectic of Sex* (1971), Ann Oakley's (1979) *Becoming a Mother* and Barbara Katz Rothman (1982) *In Labor.* Later publications too, such as Robbie Davis-Floyd's (1992) *Birth as an American Rite of Passage* and Shelia Kitzinger's (1995) *Ourselves as Mothers* have helped define the academic territory of mothers and mothering. Instead, I have outlined research carried out mainly by feminist, social, cultural and critical *geographers* on the social and cultural constructions of space and place in maternities. I have focused on this work because it helps facilitate an understanding of how mothers' identities and power are refracted through a range of psychoanalytic, discursive and material spaces. I cannot, however, imagine writing this book without having first explored some of the foundation texts on mothering (for a useful review of feminist debates on motherhood since the 1840s — sometimes referred to as first-wave feminism — see Chase and Rogers 2001: 3–29).

As mentioned in the preface and at the beginning of this chapter, the aim of this book is to open up a series of windows on maternities. However, some contemporary and highly contentious issues to do with maternities which have had a high profile in the media (and to some degree in the academy) over the past decade are not covered in the book. I am thinking of issues such as China's One Child Policy, the forced and coerced sterilization of Romani women in Slovakia, female infanticide and the aborting of female fetuses in India, and the social impacts of new reproductive technologies such as amniocentesis, genetic testing and in vitro fertilizations. These weighty topics are undoubtedly worthy of further exploration, especially using a feminist geographical poststructuralist perspective, but I have not addressed them here. Instead I have attempted to offer some windows on maternities that tend to be less frequently looked through. It is acknowledged though that there are numerous additional important windows that merit the attention of geographers and others interested in thinking about maternal bodies in relation to spaces and places.

CHAPTER OUTLINE

In this introduction I've argued that maternal bodies are bodies that conceive, become pregnant, breastfeed and nurture but that they are also much more than this. The research is informed by a version of feminist poststructuralist theory that holds on firmly to materiality. Maternal bodies are an interface between nature and culture, biology and the social, materiality and discourse. They are always located in time and space, that is, they have histories and geographies. It is imperative to deconstruct the binary between maternities and paternities, to open up material and discursive spaces to a variety of configurations of mothering in a variety of different sites. It is important to decouple maternity and women's bodies not in an attempt to evacuate mothers as the central subjects in maternity but in an attempt to 'de-essentialize' mothering and increasingly extend nurturing roles to men. Such a move is not aimed at nullifying women's 'difference' and the special value of pregnancy, birth and lactation but at recognizing the layering of meanings attached to maternity. Some men want to mother, some women do not.

Chapter 2 focuses on some of the challenges women face when they 'come out' as pregnant to co-workers, colleagues, and employers. The early part of the maternal journey is often revealed only to the most intimate of friends and family and is kept secret at work. The chapter begins with a story about my own coming out at work. It also draws on interview data collected from 19 first-time pregnant women in Hamilton, New Zealand, who discuss their experiences of telling co-workers, colleagues, and employers that they are pregnant. The chapter concludes with an argument that coming out of the closet is a useful metaphor for deepening understanding of women's experiences of disclosing their pregnancy at work and that it is difficult to generalize about pregnant women coming out as pregnant at work. Women 'do' pregnancy in many different ways, in many different workplaces.

Chapter 3 examines the ways in which pregnant women in the West use clothing as a means of constructing a range of complex and seemingly contradictory gendered subjectivities in public spaces. Like the previous chapter, this chapter draws on interview data collected from 19 first-time pregnant women in Hamilton, New Zealand. These women were asked about maternity wear, body image, fashion, activities they had continued, reduced or stopped during pregnancy, and the places/spaces they occupied during pregnancy. The chapter focuses on four different 'looks' and subjectivities that pregnant women in this research tried on: the thrifty, self-sacrificing mother to be; the sexy, proud pregnant woman; the growing woman who fears her body will be read as fat; and the pregnant professional. For first-time pregnant women making the transition to motherhood clothing the body can be a complex act. What women wear during pregnancy speaks volumes about their subjectivities — what they reveal, what they conceal, what images they create, for whom and where.

Chapter 4 builds on the work of geographers and others who have examined illness, impairment and disability. It examines the experiences of pregnant women with physical disabilities. Women with physical disabilities are often as fertile as women without disabilities but disabled women who decide to get pregnant sometimes encounter opposition, discrimination and skepticism. These women often face 'body troubles' (Dyck 1999) because dominant discourses dictate that their bodies — bodies that are considered unreliable, unsightly, sick, disabled — should not reproduce. Over the past few years an interesting literature on disabled parents (and their children who often take care of them) has emerged but still very little has been published on disabled pregnant women. The chapter concludes that pregnant women with physical disabilities inhabit some of the same emotional, discursive and material spaces as non-disabled pregnant women but their experiences can also be appreciably different from those of non-disabled pregnant women.

The aim of Chapter 5 is to illustrate in relation to maternal bodies that the moral boundary between what is considered 'normal' and what is considered 'perverse' is constantly struggled over and is spatially specific. In order to tease out this argument I offer a critical reading of approximately 20 media reports published in October and November 2002 after a New Zealand *60 Minutes* television documentary featured a pregnant woman, known as Nikki, who planned to be filmed giving birth for a pornographic movie. The chapter is informed by recent work on 'moral geographies', and Julia Kristeva's notion of abjection. It concludes that Nikki troubles the purity and naturalness of birth by constructing the space of the birthing ward as explicitly sexual. By not complying with 'taken for granted', 'common-sense' understandings of birth and motherhood Nikki opens up for question what counts as moral and shows how this is infused with geographical notions of space, place, access and boundaries.

In Chapter 6 it is argued that the notion of home in discourses of home birth needs further deconstruction. The chapter tells a story about Angeline Greensill's experiences of birthing at home. Angeline's story is placed within the historical context of birthing for Māori women in Aotearoa, New Zealand more generally. It becomes evident in the chapter that a complex (post)colonial and gender politics have inhabited, and continue to inhabit, the space of home for Māori women. While home can be an empowering space for some women to birth, however, this is not necessarily the case for all women. Home birth advocates, in an attempt to underscore the benefits of home birth, often reify home as an idealized site. Home birth is portrayed as a positive choice that offers *all* women a range of benefits. Many assumptions tend to be made about a mother's home environment, that it is private, safe, clean, comfortable, and loving. The chapter concludes that the notion of home in discourses of home birth needs to be held up to further critical interrogation.

Chapter 7 is inspired by Fiona Giles's work on 'queer breastfeeding'. Her work enabled me to begin to think through some of the spaces — material, discursive, and imaginary — of breastfeeding. Lactation has long been a symbol of maternity — sustenance, warmth, goodness — and yet there is a vast and diverse range of lived experiences of lactation, some of which reiterate dominant discourses of the 'purity of mothers' milk' and others that contest these discourses. The chapter discusses what might be considered some of the more unusual experiences of lactation in order to understand more about the spatiality of lactation — where and when breasts are understood to be in place and out of place.

Chapter 8 discusses women who are commonly understood to be 'bad' mothers. While some women are constructed as 'good' mothers (usually white, affluent women in their mid-20s to mid-30s who have a 'house and a husband'), others are constructed as 'bad' mothers (usually non-white, poor women who are thought to be too young or too old and have neither a house nor a husband). These mothers are often presumed to be lacking in some ways. This chapter focuses on three examples of mothers who are widely seen to be lacking: mothers on Welfare (who are widely assumed to be lacking financial resources), teenage mothers (who are widely assumed to be lacking in maturity and wisdom), and lesbian mothers (who are widely assumed to be lacking in sexual morals and/or being part of a 'proper' family). These women do not enjoy a heightened status when they take on mothering roles, rather they tend to be denigrated at every turn. The chapter examines discourses about 'bad' mothers and the ways in which they reflect and reinforce existing sexual, racial, and class inequalities within particular spaces.

The penultimate chapter, Chapter 9, addresses the rise of information on maternity on the Internet or in cyberspace. It heeds Rob Kitchin's (1998a: 403) appeal to examine further the complex and mobile spatialities of cyberspace. Electronic information flows have the potential to prompt social and cultural change. For example, chat rooms for women who are attempting to conceive, or women who are experiencing difficulties breastfeeding, can create a sense of belonging to 'communities without propinquity' (Webber

1963: 23 cited in Valentine 2001: 118). This chapter builds on the study by Madge and O'Connor (2005: 83) on the role on the Internet for new mothers. It also follows in the footsteps of Hester Parr (2002) who examines the rise of health and medical information on the Internet. Electronic information flows have the potential to prompt social and cultural change. The chapter examines the ways in which the Internet affects women's embodied experiences of maternity and vice versa.

The final chapter draws together some thoughts on the previous chapters in order to illustrate that maternal bodies and spaces are neither autonomous nor coherent. An argument is made to recognize the deeply contradictory practices that surround the subjectivities and spaces of mothering. Too often motherhood is seen within an essentialist and narrow framework. It is time to recognize different kinds of mothering that happen within a range of different spaces. Distinctions between mothers, fathers, and those who are considered 'childless' in relation to the care of children need to be reconsidered. It is important to decouple maternity and women's bodies not in an attempt to evacuate mothers as the central subjects in maternity but an attempt to de-essentialize mothering and increasingly extend nurturing roles to men and others.

2 'Mum's the Word'

'Coming Out' as Pregnant at Work

> You're pregnant and you're riding a wave of happiness like no other. Your family knows. Your most intimate friends know, but no one at work does. You're probably starting to think about when and how to announce your big news to your immediate supervisor and co-workers. Each pregnancy is different. The relationship you have with your boss and co-workers is different from anyone else's. So, the 'right time' to announce your pregnancy at work is unique to you. Below are some tips to help you put together a plan to announce your pregnancy at work. (Clearblue, 2006, available <http://www.clearplan.com/TellingYourBossYou'rePregnant.cfm>, accessed 22 February 2006)

> Just found out you are pregnant? Afraid to tell your boss? Learn about the best way to break the news. And, know your rights so you can relax a bit more and enjoy your pregnancy — even at work. (Pregnancy-info.net, 2006, available <http://www.pregnancy-info.net/pregnancy_and_work.html>, accessed 22 February 2006)

These quotes are from websites that offer advice on how to tell an employer that you are pregnant. They are typical of many that can be found on the Internet. The academic literature has much less to say on this topic. Extensive research has been carried out on mothers and waged work, the feminisation of the workforce, gendered organizational cultures, pregnant women's experiences of waged work, maternity leave, discrimination issues, health and pregnancy at work, and returning to work after maternity leave but there is still little published research on announcing one's pregnancy at work.[1] The aim of this chapter therefore is to address this topic. I am especially interested in some of the difficulties that pregnant women potentially face when they announce their pregnancy. It is not always easy negotiating the subject position of Pregnant Woman at work. The way I have chosen to examine this topic is to use the metaphor of 'coming out of the closet'.

Initially the closet may seem like an unusual, and perhaps even inappropriate, choice of spatial metaphors to use to discuss the experiences of pregnant women. After all, when women become pregnant, aren't they doing the very thing women are revered for — performing the sacred duty of motherhood? And yet, for some women, in some spatial contexts (especially the workplace), announcing their pregnancy to friends, family, loved ones, colleagues, and acquaintances can feel painful, risky and even shameful. Possible reasons why a woman might want to hide her pregnancy, why she might want to go unseen and remain closeted for as long as possible, are that she feels too young, too old, has no partner, no desire to be a mother, is an addict, a sex worker, a lesbian, committed to a profession, might be thought to be overly emotional or forgetful or afraid of miscarrying and feeling like a 'failure'. Fear of co-workers' and employers' responses to 'the news' can also lead pregnant women to remain in the closet at work for as long as possible even though they may be out in other places. Sometimes women disclose their pregnancy to some people in some spaces but not in others. People are told at different times. Some may be asked to keep it a secret, to keep quiet — 'Mum's the word' — to say nothing.[2]

During the past 15 years that I have been researching pregnancy it has become apparent that some women experience a great deal of anxiety and nervousness deciding how and when to broach the subject of pregnancy with their employer and co-workers. Many times pregnant women are worried about how the news of their pregnancy will be accepted at work by their boss and co-workers and what their options are. Some women and men also face telling their employers and co-workers that they plan to foster or adopt a baby or child. This issue also deserves attention but this is not possible within the space of this chapter. Here I deal only with women who are pregnant rather than with women and men who are fostering or adopting a baby or child.

The chapter begins with a brief discussion of 'the closet'. Second is a story about my own coming out as pregnant to my Chair of Department and colleagues at the University of Waikato. The third section offers some background information on the participants involved in this research. Fourth is a discussion of pregnant women's fears about harassment, discrimination and even dismissal as a result of coming out as pregnant at work. Following this the chapter examines women's positive experiences of telling employers and co-workers that they are pregnant. Two points conclude the chapter: first, that coming out of the closet is a useful metaphor for deepening understanding of women's experiences of disclosing their pregnancy at work; and second, that it is difficult to generalize about pregnant women coming out as pregnant at work because women 'do' pregnancy in different ways, in different workplaces. This mutually constitutive relationship between pregnant women and workplaces is infinitely complex.

THINKING THROUGH 'THE CLOSET'

Michael Brown (2000) in his book *Closet Space* argues 'the closet is not just a metaphor for the concealment, erasure, or ignorance of gays' sexualities.' The closet also has a materiality and exists at a variety of spatial scales, from the body to the globe, and in an array of different spaces and places. Brown (2000) explains that the closet is a key term in queer theory and parlance because it relays so effectively the specificity of subjugation based on sexuality and gender.

> It [the closet] allows us to speak our anger and pain about lying, hiding, being silenced, and going unseen. The closet's ontological demands are exacting and exhaustive; that we cannot be in the world unless we are something we are not. ... it also signifies the inevitable oppression we face if we 'come out of the closet' either by choice or force. (Brown 2000: 1)

Brown (2000: 147) notes 'People can be in and out of the closet simultaneously through resituating that broader environment. Its space can reveal and conceal at the same time, often dependent on one's own location.' Diana Fuss (1991) makes a similar argument when she says to speak of being in or out of the closet is problematic because it creates a binary opposition which tends to construct being in the closet as dark and oppressive and being out of the closet as light and liberating (also see Butler 1991; Sedgwick 1990). Fuss (1991) explains the difficulty with the inside/outside rhetoric is that it disguises the fact that many people are both inside and outside simultaneously.

Despite these criticisms coming out of the closet can be a useful metaphor for understanding a variety of different embodied subjectivities that people, for one reason or another, might feel troubled about disclosing. Over the past few years the notion of coming out has become used increasingly to describe a variety of disclosures (not just sexual) including coming out as an alcoholic, coming out as ill, or even coming out as a Christian in secular circles or as a conservative in liberal circles. It seems there are many closets. Coming out of one can mean entering another. For example, after I came out as pregnant I went into the closet to drink wine (there is growing 'public concern' about fetal alcohol syndrome). As a pregnant woman I felt under surveillance in public places such as pubs, bars and clubs and therefore only ever drank non-alcoholic drinks in these places. At home, though, I occasionally enjoyed a glass of wine.

The notion of the closet has been developed under the guise of queer theory. The term queer, as a critical theory and methodology, originates in gay and lesbian studies. Queer has provided a platform from which to critique heteronormativity (Sullivan 2003). Eve Kosofsky Sedgwick argues, however, that queer theory can be usefully applied to a broader range of normative knowledges and identities than just sexual ones. She explains:

> a lot of the most exciting work around 'queer' spins the term outward along dimensions that can't be subsumed under gender and sexuality at all: the ways that race, ethnicity, postcolonial nationality criss-cross

> with these *and other* identity-constituting, identity-fracturing discourses, for example. (Sedgwick 1993: 8–9, italics in original)

Giles (2004), interestingly, yokes together breastfeeding with this definition of queer as 'identity-fracturing' by pointing to the merging of mother and infant subjectivities. Giles (2004: 303) argues 'If it is accepted that the normative Western subject is developmentally reliant on individuation, then breastfeeding, however conventionally practiced, is perhaps already an activity which "queers" us.'

When women tell people at work they are pregnant their identity is fractured. The individuated worker becomes a maternal subject who is no longer one but two (or more). Her corporeal boundaries are radically reconfigured as her body grows to accommodate the baby growing inside her. Giles (2004: 303) notes that there is likely to be some resistance to queering something that is not usually perceived to be overtly sexual but it could be argued that 'queering can usefully be applied to any behaviour that has been extravagantly regulated, especially where those regulations have repressed any sexuality that the behaviour might once have owned.' Pregnancy brings sex into the workplace. Arguably it also queers the workplace.

'WRITING ONE'S LIFE': A PREGNANT GEOGRAPHER

On disclosing my own pregnancies at the University I felt like I was bringing sex into the workplace, like I was queering the hallowed halls of academia. Pamela Moss (2001: 7) notes 'Autobiography and geography are not strangers. Historically, autobiography has been used to chronicle geography as a discipline'. Moss suggests there are many benefits of 'writing one's life' to the continued construction of the discipline of geography. Following Moss's lead I reflect here on my own coming out as pregnant in 1994 to my chairperson and colleagues in the Geography Department at the University of Waikato. I hasten to add that I am not the first geographer to write about coming out as pregnant, or at least coming out as a mother. Carol Ekinsmyth, Rebecca Elmhirst, Sarah Holloway and Helen Jarvis (2004: 95) write their autobiographies as mothers explaining:

> We recognize the very real dangers of 'coming out' as mothers and admitting — dare we? — that our children come first. How regrettable, alien and disabling it is that still, in the 21st century, we are acutely aware that love and family responsibilities are incompatible with hegemonic constructions of paid employment (and even academia) that demand total commitment.

In 1994 I was already out as pregnant to my family and friends but in my mind my chairperson and colleagues posed a difficulty. My partner and I decided to attempt to have a second baby. I had been working as a lecturer in the Geography Department at Waikato for two years. My first child was four. Despite the Department having a reputation for supporting 'others' such as women, lesbians, and Māori I had a strong sense that having a baby would potentially disrupt the teaching of courses, hinder my research, and risk coding me to colleagues and students as being well and truly 'on the mommy track' (as it is sometimes referred to in the United States). I was already juggling motherhood and career (see Ekinsmyth *et al.* 2004: 100), but this juggling was about to become more intense. I was not concerned about being unmarried but I was concerned that as a pregnant woman I would not be taken seriously in the academy.

There were four other academic women employed in the Department at the time but only one had children and her children had grown up and left home. This colleague had nevertheless to some degree paved the way. I carefully planned the conception (as much as one can plan these things) so that the baby would be due in the break between semesters so as to minimize possible disruption (when I reflect on this decision now I am surprised that I was so willing, in a sense, to prioritize the institutions' needs over my own). Everything went to plan. At approximately four months pregnant, when the Department met to organize teaching for the upcoming semester, I 'announced' that I was pregnant but hopefully would be able to return from giving birth straight after the semester break to take up a full teaching load. I remember feeling very nervous and apprehensive about having to tell colleagues this even though I was clearly attempting to be (almost unbelievably) accommodating.

As it turned out my colleagues seemed happy that I was pregnant and perhaps relieved that I would be taking only very limited parental leave. I still vividly recall that meeting. Disclosing my pregnancy felt like a coming out of sorts. I gave out more personal information to my colleagues than I felt comfortable with. I felt a certain pleasure and delight that I was having a baby — proud that my body was able to do this amazing thing — but I also felt shame. I felt shame that my colleagues knew I'd had sex but I felt even more shame that maybe I was not a 'serious' academic after all (see Johnston 2007 on lesbians feeling both pride and shame coming out at gay pride parades).

Most of the senior women geographers I knew and admired — my role models — did not have children, or at least if they did, they did not make public this part of their lives. They had entered the academy at a time when there were still huge constraints on women and many I suspect felt they needed to forgo marriage and motherhood. These women, it seemed to me, were entirely committed to their work and I admired them for it. I too wanted to be productive at work (not always juggling) but I worried that having a second child would prohibit this. It seemed I could still 'get by' doing two jobs — being a mother and an academic — with one child but I wondered if my familial responsibilities with two children would impossibly curb my professional opportunities. I was aware of the structural conditions that would impact upon my attempts to combine academic work and motherhood and yet still I chose to get pregnant for a second time (see Monk 2001 on 'the personal and professional lives of women geographers'). In my less confident moments I thought perhaps I had made a foolish decision and that women in my generation who grew up in the 1960s and 1970s could not 'do it all' and 'have it all' after all!

My experience of coming out as pregnant to my chairperson and colleagues at the University made me wonder about other women's experiences in other workplaces. How similar or different were they to my own? In the next section of this chapter interview data are presented from a study of some first-time[3] pregnant women who live in Hamilton, New Zealand.[4]

PARTICIPANTS IN THE STUDY

The data presented in this chapter, and the following chapter, were derived from 19 first-time pregnant women who were recruited from Hamilton, New Zealand to participate in the project (see Figure 2.1). These participants responded to an article on the proposed research published in the local newspaper (*Waikato Times*, 6 November 2002). The newspaper article (which was accompanied by photographs of four pregnant celebrities — 'singer Celine Dion, Italian glamour model Nina Moric, girl-band singer Nicole Appleton and actor Demi Moore') called for participants explaining that I had conducted research ten years ago on first-time pregnant women in Hamilton and found that many women hid their 'bump' in public or withdrew from public places altogether. The article explained that I was keen to ascertain whether celebrities parading their pregnancy in public had changed Hamilton women's attitudes towards revealing their pregnancy. The questions did not focus solely on women's experiences of telling their employers and co-workers that they are pregnant (the focus of this chapter) or maternity wear and body image (the focus of Chapter 3) but covered a range of topics. For example participants were also asked about the activities they had continued, reduced or stopped during pregnancy and places and spaces they occupied during pregnancy.

Of the 19 pregnant women who were interviewed, one was at secondary school, one had recently quit her job as a café worker, and the other 17 were still employed in a variety of occupations including two primary school teachers, natural therapist, occupational therapist, dairy (small local grocery store) sales assistant, marketing communications manager, public sector (government) scientist, university lecturer, radio announcer, surgical claims insurance assessor, personal assistant, office worker, insurance com-

Figure 2.1 Photomontage of Hamilton, New Zealand. *Source:* Photographs by Betty-Ann Kamp.

pany claims clerk, home loans consultant, registered nurse, street performer and family support worker (see Appendix for more details about the methods used for data collection and analysis for this chapter and for a list of the participants who were interviewed). In addition to drawing on this interview data I also draw on relevant information on disclosing one's pregnancy at work from various websites, women's magazines and daily newspapers.

DISCRIMINATION AND DISMISSAL

It is difficult, if not impossible, to ascertain exactly how many pregnant women in various countries around the world engage in paid employment. In New Zealand (where I live) there is no statistical information on the number of pregnant women who work in paid employment each year (Collins 1995). It is possible, however, to make an educated guess. Sharon Collins (1995: 7) suggests 'Over 14,000 women in the New Zealand workplace are likely to become pregnant in any one year.' While this is only a general indication it demonstrates that pregnancy at work is an important issue for many working women (see Brown 2000 on employment during pregnancy). In any given year there are likely to be many thousands of women confronting the issue of telling their boss and co-workers that they are expecting a baby. While for some women this may be a pleasure, for others it is problematic.

At least half of the participants in this study were apprehensive about announcing their pregnancy at work partly because they feared exposing personal information about themselves and partly because they feared discrimination, and even worse, dismissal. Olivia, a Nursery Plant Sales Assistant, aged 32, decided to 'pluck up the courage' to tell her boss quite early on that she was pregnant because the work she was doing was physically strenuous and she wanted to reduce the number of hours worked. Olivia's job in 'dispatch' involved her receiving orders, riding a motorbike and a trailer around the nursery, picking up heavy potted plants and trees, putting them in the trailer and unloading them. Olivia explains:

> When I initially told him ['the boss'] I was pregnant, he said 'oh, that's fine, that's really good.' But then he said another girl that was pregnant, in dispatch, worked right up until she had the baby but I found out later that she didn't. Like, he thought that she did, but it was a while ago and he just thought that she did but she didn't. And then I asked to take Fridays off because I was just so tired. I'd just come home from work and just cry because I was so absolutely knackered. And so I said 'I've just got to take Fridays off' and they said that was okay, but, like even, because we have to get the trucks out like on the, say Tuesday, Wednesday nights, sometimes it would be overtime. And I'd

still like be there, they wouldn't come and say to me, you can go home at four thirty, I'd just have to stay on and do it.

Olivia felt pressure from her boss to work right up until her due date because this is what another 'girl' at work did. She also felt pressure from her co-workers to work overtime even though she found the work to be extremely exhausting now that she was pregnant (especially towards the end of her pregnancy). Olivia experienced increasing problems with her sciatic nerve and was not able to keep doing heavy lifting but she felt pressure to keep working the same number of hours:

> It's just that we were just getting busier and busier at work as well, you know, like spring and bedding the plants in. And I just, also my stubbornness as well, you know, for the money and stuff like that too, but, you know you feel like you don't want to let the team down. Same with going home at four thirty, I felt I couldn't because I've got to leave my friends working.

Olivia, in order to keep doing her job, needed some special consideration. Her job was strenuous and so she needed to be able to work fewer hours but then felt by doing so she would let down 'the team' and her 'friends'.

In New Zealand the Human Rights Act 1993 covers women against unlawful discrimination on the grounds of sex and this includes the 'condition' of pregnancy. Despite some dismantling of barriers to equal participation in paid employment for pregnant women and despite it being illegal under the Human Rights Act to discriminate against a worker on the grounds of pregnancy, pregnant women continue to be treated as unwelcome intruders and burdens in some workplaces. The normative subject in most workplaces is an individuated subject. When women become pregnant, however, the entity they call the 'self' fractures and the individuated subject becomes at least two.

Like Olivia, Jennifer, a 32 year old scientist, was also worried about letting down the team. She thought that she would be treated differently once she told her boss and co-workers that she was pregnant. Jennifer says:

> With work it was extremely difficult in the early stages — up until probably 18 weeks, until I started to show, because it was my secret. And I was really scared that I'd be treated differently, such as 'oh, she's on her way out, she's leaving, her attitude to work has kind of changed, or we won't give her interesting projects because she's only got x amount of months [left]'. Basically being written off is what I was afraid of.

Jennifer's workplace is highly masculinist and she was afraid of being thought to be no longer completely committed to her job (see Massey 1996 on gender constructions in the UK's 'high-tech' industry). Jennifer was finding work a 'struggle' but did not feel comfortable sharing this information with anyone because she feared that it might undermine her reputation for hard work. In a sense Jennifer feared that her pregnant body might queer her workplace and her, and therefore she remained in the closet until her pregnancy began to show. Jennifer says:

> You don't want anyone to know how much you're struggling through the day, because you don't want to be observed, especially by the males, you don't want them seeing you, thinking 'oh well, oh yeah, she's slacking off' or ... when you've always worked long hours and gone the extra mile you don't want the attitudes to change towards you.

It can be particularly difficult for women in highly masculinist workplaces to tell their employer and co-workers they are pregnant. Linda McDowell and Gill Court (1994: 737) in a discussion of merchant banking in London refer to bodily imagery such as 'having the balls to pull off a deal' (also see McDowell 1995, 1997). They also make reference to the successful traders who are sometimes referred to as 'big swinging dicks'. This leaves *all* women in highly masculinist workplaces in a complex and difficult subject position but during pregnancy women are even more likely to find themselves caught in a difficult position because it forces something of an end to the 'masculine masquerade' (McDowell 1997: 197). McDowell (1997: 197) argues that some women, especially younger rather than older women, attempt a 'masculine masquerade'. Pregnant women tend to be coded as 'more feminine' than women who are not pregnant thereby heightening the challenge of being an 'honorary man'. Pregnancy is the ultimate marker of femininity. McDowell (1997) argues that being an honorary man is doomed to failure. Being pregnant disrupts even further women's attempts to 'be one of the boys' because it draws attention to women's female gendered and sexed embodiment.

This notion is reiterated in Faye Brown's story reported on the Internet (see Pinky-mychild.com, 2005, available <http://www.pinky-mychild.com/features/pregnancy/keeping.html>, accessed 20 June 2005). Brown is a mother of three and director of a clothing label — Motto. She felt that she needed to stay closeted at work as long as possible. Brown speculates that the perception of pregnancy as fatness can have repercussions for a woman's corporate image (see Earle 2003; and Chapter 3 this volume). She explains 'from our culture and from advertising, there are misconceptions that "fat equals lazy" and breasts are equated with homeliness, rather than efficiency.' Brown concealed her first pregnancy for as long as possible. She says:

> I didn't want to be seen as not being capable, but it was more about where I came from. A lot of women have had to think like a man to make it. Working alongside men, we've taken on the persona that says,

> you have to have balls to make it with the boys — and where does that leave pregnancy? (Pinky-mychild.com, 2005, available <http://www.pinky-mychild.com/features/pregnancy/keeping.html>, accessed 20 June 2005)

Joy, a 27 year old office worker who I interviewed took maternity leave at 34 weeks pregnant. Joy experienced a miscarriage with her first pregnancy and so did not tell people, including her boss and co-workers, that she was pregnant until the second trimester when she was beginning 'to show'. She explains: 'At about twelve weeks we started telling people. We kept it a lot more of a secret this time round so that there weren't so many people disappointed if it [miscarriage] happened again.' By the third trimester of Joy's pregnancy she began to feel that she was less valuable to her employer because of a reduction in her attention span. She expresses concerns about becoming a burden in the workplace rather than a productive worker. Joy says:

> I've found my attention span has dropped dramatically. I just find myself, when I was at work it was like 'come on, I've got to finish work because otherwise I'm going to be so useless to this company!' [laughter] They're going to sit there and go 'why are we paying you?'

In this instance it was not Joy's employer or co-workers who judged her performance at work but rather Joy herself who judged her own behavior (see Foucault 1977 on self-surveillance).

Ainsley, a performance artist, claims that when she told her co-workers she was pregnant 'they thought I was joking [chuckles]'. She continues:

> I said 'look, it's not going to change my job at all, I won't let it' and yeah, that's it basically. No one has ever sort of gone 'oh but you'll need time off and you'll need to do this now that you're pregnant.'

However, being pregnant did change Ainsley's job. Contradicting her earlier statement later in the interview, Ainsley says that since becoming pregnant her centre of balance has changed and she now feels more clumsy and has therefore stopped stilt-walking. She adds: 'besides, I'd look bloody stupid up on stilts trying to wear some of those skimpy little tight outfits'. Ainsley, unlike Joy, was not concerned about her changing role at work. She easily took on the role of 'crew instead of cast for a change' and made costumes, organized events and so on. She says 'my not being able to perform leaves more work open for them really.'

Interestingly, all but one of the participants noted some change in their mental capacities since becoming pregnant. Pregnant women are widely thought to exhibit behaviours which could be considered 'hysterical', in particular, becoming 'overly' emotional, not being able to think rationally and frequently forgetting things (Longhurst 1997). This is not to say that women 'really' do become more emotional, irrational and/or forgetful during pregnancy (it is not some kind of biological given or truth) but neither is it to say that it is just something that is said by or about pregnant women. Rather, changes in body image and material form come from social, psychical and material sources. Body image and material form are inseparable and mutually constituted.

Perhaps more importantly it needs to be noted that hegemonic discourses function to keep in place particular societal arrangements. The discourse that pregnant women tend to be 'overly' emotional, irrational and forgetful functions to discourage and/or exclude women from certain spaces, such as some work spaces, on account of their supposedly 'natural' 'condition'. Historically, people deemed incapable of engaging in rational public discourse such as indigenous people, women, and those who are intellectually and/or physically disabled have been excluded (and to some degree still are) from spaces of power.

> Such groups were (are) considered too bound to their bodies, too dissipated in their intellect to be able to engage in independent thought and action. Their 'differences' disqualify them from stepping 'objectively' and 'dispassionately' into the public sphere and engaging in 'public affairs'. (Longhurst 1997: 38)

The discourse of women becoming overly emotional, irrational and/or forgetful during pregnancy can be particularly problematic for pregnant women engaged in (paid) work because workers are usually expected to function as fully individuated and rational subjects who have consummate control over their mental functions.

Ainsley says that her brain is:

> just fucking mush — forgetful and clumsy and just mush really ... I go through insane periods where I just can't remember people's names ... just completely blank. I feel like a complete dickhead.

Nia worked as a natural therapist. She says, 'I've suddenly gone to mega slow mode, and I think really slow, and talk really slow, and I forget words, names, everything'.

Diana, a marketing communications manager, aged 32, when asked about her capacities at work, tries to articulate the complexity of separating her 'true' feelings from culturally constructed notions of pregnancy. She states, 'I don't feel stupider, but I have, I do forget a lot of stuff. I thought it was just an Old Wives Tale, but it's true!'

Women continue to be discursively constructed as forgetful and having reduced mental capacities during pregnancy. These discourses of forgetfulness are inextricably entwined with the physiological changes of pregnancy.

Some medical research goes so far as to postulate that the brain actually shrinks during pregnancy.

> If a pregnant woman is easily forgetful or has difficulty concentrating, British researchers think they know the reason why — the brain has a tendency to shrink late in pregnancy, suggests a report in New Scientist. ... It may take up to six months for women to regain the loss in brain size, according to Anita Holdcroft, an anaesthetist at the Royal Postgraduate Medical School in London. (Pregnancy shrinks brain, 2006, available <http://www.personalmd.com/news/a1997011304.shtml>, accessed 29 September 2006)

It is not only medical discourse that reinforces the notion of pregnant women as forgetful but also parenting discourses (see, for example, A healthy me: forgetfulness, 2000, available http://www.ahealthyme.com/article/bcrefcap/100322064, accessed 10 August 2005).

Being constructed as lacking in mental capacity can be especially problematic for women who are expected to be highly intelligent and articulate, for example, university lecturers. Carla, a university lecturer aged 43, was 30 weeks pregnant at the time of interview. She worked up until a few weeks before her due date but felt the need to keep conveying a professional image at work. She was concerned that being pregnant might be read by some, mainly her students, as being unprofessional. Carla carefully 'managed' how she dressed so as to not reveal her pregnancy to students for as long as possible. She claims:

> I managed it — I mean I was much more careful about my appearance at work than outside of work ... just because I mean when I was teaching I just didn't want students sort of fixating on the fact that I was pregnant ... I thought, and I said, 'yeah well I was worried that students especially would have a different reaction' and so most of them I don't think were aware that I was pregnant until after the semester ended, so I was really careful in that regard. It wasn't like I cared that much in terms of other people in the department, I was really more worried about maintaining this professional image ... I was much more careful about it at work, and then I'd be home on the weekend and I'd be going out dressed in fitting clothes and think 'oh what if some of my students are there, they're going to know I'm pregnant.'

Denise, aged 29, worked as a personal assistant for a senior woman. Denise decided to tell her employer before any of her colleagues so that she did not hear the news through the office grapevine (common advice offered on many Internet sites is tell your employer the news directly because this will appear more professional than him or her hearing it from a co-worker). Denise felt that her employer would appreciate knowing early on to allow as much time as possible to make arrangements for her absence. Denise's employer took the news well and was supportive.

This was not the case, however, for Diana who worked as a Marketing Communications Manager. Diana found her colleagues to be generally supportive but her employers were not. They attempted to terminate her employment. Diana explains:

> Generally work people are, it's quite cool because, you know 'you're pregnant, its really great!' But yeah ... because my position's a key one they [Diana's employers] wanted to replace me full-time. I still wanted to work during my pregnancy. I still wanted to go on maternity leave and they said 'you can't, we have to terminate your position'. And I said 'you can't say that to me!' That's destructive dismissal. I just went and found out exactly what my rights are and told them 'you can't do that, it's destructive dismissal, if I meet the criteria, which I do, you have to allow me maternity leave' and so ... the marketing manager couldn't replace me full-time, she had to find a part-time replacement, so then I said 'no you don't, that's not going to suit either' so we sorted it out in the end.

Diana experienced what is probably one of the worst possible outcomes when she came out to her employers as pregnant. They attempted to terminate her position.

Lana, aged 19 and 30 weeks pregnant at the time of interview also found telling her employer difficult. Her boss also tried to terminate her position. Lana explains in detail:

> I was working in a café but I had an argument with them over being pregnant, so I just walked out. Because I worked there all year, and they like, you know, we got on well, then they wanted to take me further, like they took me on a coffee training evening and then they wanted me to go to a foods based evening and things like that. But when I told them I was pregnant and I would be leaving work just before I was due they didn't like that and they said 'well, you've got to get rid of it'. And like, I don't believe in that, so I thought, you know, I've just got to put up with whatever comes, you know? They made, they came up with, like, scenarios like you're too young, you're not going to be able to cope, no-one's going to help you ... because at that time I wasn't on talking terms with my parents, so I didn't have their support. But then, I thought that I'd rather have their support and no job than a job and no support so I quit my job and I went back home. So I'm just on a [unemployment] benefit now.

Lana's employer's response 'you have to get rid of it' seems inappropriate. Even though Lana appeared to have been doing well in her job, her employer and co-workers (she says 'when I told *them* I was pregnant') saw fit to offer unwelcome advice. Lana sought other employment but was unsuccessful.

> I tried getting a new job, but no, you can't get a job anywhere, like, I am so bored now, like, because I haven't worked for so long ... nobody said it, but they just said, 'oh, no there's no work here, you know'. Because you know, I tried like almost every little town around where I am living — even Pak'n'Save, and Woolworth's [chain supermarkets] and all those jobs, but it was just 'no' so I just worked with my dad for the while, just on the farm, because we live on a farm, and then just get the benefit [welfare].

As a reasonably 'young mum' (see Chapter 8) Lana troubled the discourses of 'individuated worker' *and* 'good mother'. Consequently she was dismissed — metaphorically and literally.

Camille Guy (1992: 33) reports in the *New Zealand Herald* that women often report that they have been dismissed once their employer learns they are pregnant.

> This dismissal of pregnant women is despite the fact that in a major study conducted in the United States on four states by the Families and Work Institute in 1988 it was found that 'the vast majority' of supervisors noted that pregnancy had no adverse effects on job performance. (Guy 1992: 33)

It is near impossible for women to find temporary jobs for the remainder of their pregnancies.

Another participant who also considered herself to be a 'young mum', Mihi, is of Samoan origin and aged 17. She was still at school when she became pregnant and decided to remain closeted to all except some close friends and a few teachers. While teachers aren't exactly employers and fellow students aren't exactly co-workers, clearly there are some institutional parallels. Mihi figured that she would be able to 'get away with' not disclosing because by the time her pregnancy would be clearly evident she would have completed her exams and her final term. She told me:

> Nobody at the school knows, well, my close friends do but I told them not to make it public news because of the reaction. Older people understand but people my age don't. Yeah, just so I felt more comfortable, like, not being there, not showing it, and not having people question me about it ... I just haven't told people.

In many instances pregnant bodies at secondary schools and at work disrupt ideas of what it means to be a 'good' student or a 'good' employee. However, as McDowell (1997) argues so persuasively it cannot be assumed that disembodied, asexual, rational, masculinity is still a requirement in all professional occupations. McDowell explains that too often studies of gender in the workplace assume or take for granted women's inferiority.

> The ways in which multiple gendered differences become important in the culture of organisations and the ways in which these change are thus not investigated and the stories told are too undifferentiated. Instead of looking at how workplace interactions themselves gender women and men in multiple ways — some less, some more acceptable than others, some appropriate to one type of work or site of work and some, perhaps quite other ways of being female and male, to another type or site or work — the dominant model of disembodied rational masculinity that used to be so important in professional occupations is taken for granted. (McDowell 1997: 156)

To end the story of women coming out as pregnant at work here, therefore, would be amiss. Of the 19 women interviewed approximately half reported receiving positive reactions from their employer and co-workers when they announced their pregnancy. It is to these narratives that I now turn.

AFFIRMATION AND APPROVAL

At least three of the interviewees reported working in 'pregnancy friendly' workplaces. Bronwyn, an occupational therapist aged in her 30s, comments:

> They [her boss and co-workers] were really supportive and when I left they gave me all these little presents and stuff and one of the ladies that I've had some on and off times with — everyone does, she's just one of those people — she snuck in on the last day and gave me this present and I heard her saying to the boss 'Oh [Bronwyn] is leaving in two days' and she sounded quite mournful ... and she brought me this present. It was quite sweet.

Occupational therapy is a largely female-dominated occupation. So too is primary school teaching. Kate received numerous affirmations from her employer, colleagues and students in the primary school where she works. Kate explains:

> Some of them are quite neat because they give me hugs and things. One little girl, she gave me a hug and said 'we're hugging your baby!' That was really cute. But it surprised me how knowledgeable they were actually, like I took in my [ultra-sound] scan [of the baby] once, and they've obviously, a lot of them have seen their own scans. Yeah, so that's sort of neat.

Cathy also works in a female-dominated occupation. She is a registered nurse and works in an emergency department. Cathy was one of six women who were pregnant at work. It is perhaps not surprising that in female-dominated occupations, employers and co-workers are more used to working with pregnant women. Cathy notes: 'There's six of us at work that are pregnant — six girls at work — so we all have these discussions about what we can and can't do'. Cathy said her boss had been 'pretty good' when she told her she was pregnant.

> She just says you should keep up with exercise and that sort of thing because I think with my job you have to be quite fit because you're on your feet for eight and a half hours a day and you're walking around a lot and she just said to me 'if you want to keep working as long as you can, just keep fit and healthy and stimulated'.

Sometimes, however, support and advice, especially from co-workers, can extend beyond the bounds of helpfulness. Cathy, for the most part, felt supported at work claiming 'on the whole everyone's been really good' but comments 'some people I work with have tried to wrap me in cotton wool a little bit'. Joy, the previously mentioned office worker, went as far as saying that her co-workers seemed to think she was an 'incubator'.

> They [co-workers] were concerned about the baby, not about me. But they were concerned about me because I was the incubator for the baby, so yeah. It was, yeah, it was very much 'how's the baby?' and some of them didn't even ask how I was. At least my friends asked how I was.

Joy also made the point that she receives a lot of advice from co-workers. People who she had hardly even said hello to in the past suddenly sat with her in the tea-room in order to tell her stories about when they were pregnant and to offer advice. As I have argued elsewhere (Longhurst 1999), pregnancy is an important 'rite of passage' and one facet of this rite of passage is pregnant women giving, and more usually, receiving advice (women who are pregnant for the first time are especially likely to receive a great deal of advice). This advice comes from a range of people including employers and co-workers. The advice often focuses on topics such as diet, exercise, birth, medical procedures, lactation, how to care for a newborn baby and how to raise a child.

> Although this battery of advice is to an extent welcome and regarded positively, many pregnant women also experience a sense of being under surveillance, and of being regarded as vessels for a fetus whose well-being is the primary object of the advice-givers. (Longhurst 1999: 78)

Being in a female-dominated workplace doesn't necessarily guarantee that disclosing a pregnancy will be 'easy' and that just the right amount of support/advice (not too much, not too little) will be forthcoming. Nia, aged 26, works as a natural therapist, a female-dominated occupation. She was uncomfortable telling colleagues that she was pregnant not because she thought they would assume her to be less competent at her job but because she felt that being pregnant positioned her as an overtly sexual being. Nia explains, 'I felt a bit self-conscious that there was probably, there might be people thinking, like, sexual things about me, or thinking "oh, she's had sex!" or something like that.' Nia's feelings of embarrassment about disclosing her pregnancy extended beyond colleagues to family:

> Also when I had to tell my family that I was pregnant, I felt quite embarrassed, because it was pretty obvious that I would be, but, having sex and that. But, I felt kind of embarrassed to tell them, especially my granddad.

CONCLUSION

What is evident from the narratives presented in this chapter is that women's experiences of coming out as pregnant at work vary considerably depending on the type of work they do, the culture of the workplace, the relationships they have with their bosses and co-workers, their feelings about their pregnancy, and various other aspects of their subjectivity. By and large, however, it seems that coming out as pregnant in female-dominated workplaces does not cause as much 'gender trouble' or queer the workplace as coming out in highly masculinist workplaces.

Coming out of one closet can prompt a perceived need to enter another. Coming out as pregnant at work often means that pregnant women are then expected to behave in particular (maternal) ways. They may feel that they need to modify, change, or hide certain behaviors that fall short or exceed these expectations. Others monitor their behaviors but pregnant women also discipline themselves in what are seen to be appropriate ways.

Many discourses on pregnancy and workplaces are underpinned by liberalism and focus on equal rights and discrimination. For example, the BBC News in the United Kingdom on 1 September 2003 reported 'A series of pregnancy discrimination "horror" stories has prompted the first ever investigation into the way pregnant women are treated at work.' The Equal Opportunities Commission planned to launch a 17-month inquiry because it had received more complaints about the subject that any other issue. In the United States women who combine career and motherhood are said to be on the 'Mommy Track' (unlike 'career-primary women'). Guy (1992, 33) explains:

> A *Harvard Business Review* article suggests employers classify women according to which track they are on, clearing any artificial barriers to career promotion for the single-minded women and offering flexibility to the Mommy Trackers. The article sparked fierce debate, especially over who decided which women were on which track and why men were exempt from such conflicts.

No doubt such discrimination against pregnant women does exist. Some pregnant women, however, report feeling included and even affirmed in their workplaces. Women's experiences of announcing their pregnancy at work vary enormously depending on a myriad of factors including the culture of the workplace. In other words, there is no one model that fits all in relation to disclosing one's pregnancy at work.

But perhaps even more importantly, the point needs to be made that the politics of equality and inequality, inclusion and exclusion can only go so far in explaining pregnant women's experiences of workplaces. Rather than understanding pregnancy in the workplace through liberal feminist discourses, this chapter has drawn on the notion of the closet, as articulated in queer theory, to provide a different lens through which to view pregnant women's experiences in workplaces. Coming out as pregnant is a gendered performance. Women 'do gender' and 'do pregnancy' differently in different workplaces. A mode of enquiry that focuses on notions of queer has the potential to open up attitudes towards pregnancy as a bodily 'condition' that tends to be marked as different.

To state the obvious, not all workplaces are the same. Some workplaces are highly masculined, some are highly feminised, many combine elements of both. Neither workplaces nor men and women fit into neat and tidy binary categories. Instead gender is performed in multiple ways in a range of different workplaces. Sometimes women carry out feminised roles in highly masculinised workplaces and vice-versa. Workers engage in complex embodied performances that transgress gender categories. Work itself and workplace cultures are continually in flux. Although it is often the case, it cannot be taken for granted that coming out as pregnant will be difficult or nerve-racking for all women in all workplaces. In more feminised workplaces women can be revered for becoming pregnant. Elspeth Probyn (2000: 61) argues 'the lines of force that regulate and actually produce us are always in motion; [so] that the entity we call ourself is equally always in motion'. Coming out as pregnant at work marks a point when the entity we call ourself is in motion. Like for lesbians, gays, bisexuals and others who come out of the closet about their sexuality, it will more than likely simultaneously involve pride and shame (Johnston 2007).

3 (Ad)dressing Pregnant Bodies

Clothing, Fashion, Subjectivities and Spatialities

This chapter provides another window on maternities. It (ad)dresses clothing and fashion for pregnant women and the ways in which they express their subjectivities in a variety of different spaces, including workplaces.

In 2003 a new doll called Midge, went on sale in many Western countries. Midge is part of Mattel's 'Friends of Barbie' collection (see Randolph 1998 on 'That Barbie doll look'). What makes her different from other dolls currently on the market is that she is pregnant. Midge's large pregnant stomach is kept in place with small magnets. Lifting it off reveals a baby doll tucked in a fetal position. 'Midge and baby' come complete with accessories including a crib-come-changing table, diaper, and baby monitor. Like Barbie, Midge is tall, has long slender arms and legs, pointy breasts, and long blond hair. She has a gold wedding band painted on her finger (it can't be removed). Midge wears a lilac coloured V-neck dress that fits firmly over her large stomach. Under the dress she wears bikini briefs. On her feet are high heeled platform sandals (see Figure 3.1).

I begin with this story about Midge because I want to (ad)dress pregnant bodies, that is, to provide an 'embodied reading' of pregnant women's accumulation and wearing of clothing. Dressing negotiates between the personal and the social, hence clothing the body can be one of the most complicated acts of daily existence (Niessen and Brydon 1998). The chapter examines the ways in which pregnant women use clothing as a means of constructing a range of complex and seemingly contradictory gendered subjectivities in different spaces. As Niessen and Brydon assert 'No discussion of fashion and clothing can escape the centrality of gender, sexuality and desire to their meaning and use' (1998: xii). This centrality is accentuated during women's first pregnancy when they try on different clothes, take up different gendered subject positions, and inhabit different spaces.

The chapter is divided into five sections. The first section reviews recent research on clothing as a means of constructing individual subjectivities and social meanings in different spaces. Particular attention is paid to those authors who discuss maternity wear. Second, the methodological process used to carry out the research is explained. The third section charts some of the shifts over the past two decades from maternity wear to maternity

Figure 3.1 'Midge and Baby' from the 'Friends of Barbie Collection'. *Source*: Photograph by Robyn Longhurst.

fashion. The advertising and retailing industries in the 1990s began marketing products to a new group — pregnant women. The fourth, and most substantial section, presents interview data on four different 'looks' and subjectivities tried on by pregnant women in this research. The chapter concludes that clothes matter not just for pregnant women, but for others as well. Fashion and clothing are cultural constructions of embodied subjectivities which are intimately bound to spatialities.

CLOTHING, FASHION, SUBJECTIVITIES AND SPATIALITIES

Kellner (1994: 160) argues 'In modernity, fashion is an important constituent of one's identity, helping to determine how one is perceived and accepted.' Despite this, Niessen and Brydon assert, 'Fashion and clothing have for a long while remained scholarly unmentionables' (1998: ix), construed as feminine, frivolous, subjective, and unworthy of academics' attention. More recently, however, as some of the conventional barriers of academia have disintegrated and disciplinary boundaries have blurred, fashion and clothing have received much more attention.

In 1997 a specialized journal began to be published, *Fashion Theory: The Journal of Dress, Body and Culture,* which aims to provide an international and interdisciplinary forum for the critical analysis of fashion as a cultural construction of embodied identity. The term 'fashion' is difficult to define because it is simultaneously a business and an art-form. Its dimensions are economic, aesthetic, cultural, social, and psychological (see de la Haye and Wilson 1999). Fashion is often divided into high fashion (elite designer) and everyday fashion (clothing behaviour in general) but the relationship between the two is disjointed and complex. High fashion does not simply 'trickle down' to the masses (Craik 1994: ix).

An important constituent of fashion is clothing. It is vital to examine and understand clothing and clothing practices because most of the time people live their lives not just as bodies but as 'clothed bodies' (Banim *et al.* 2001) who inhabit particular spaces. Clothes are about signification and emotion. They are also about survival and materiality. Clothes get stained, faded, age, lose their shape. This is often forgotten in the plethora of semiotic and sociological analyses of fashion and clothing. Banim *et al.* (2001: 4) argue, therefore, that we need more accounts of 'clothes as lived experience'.

In this chapter I am concerned with the linkages between Western fashion, clothing, subjectivity and spatiality. I use the term 'subjectivity' (rather than 'identity' which is more commonly used in literature on fashion and clothing) because I want to highlight the 'spatial imperative' (Probyn 2003) of subjectivity. Subjects are interpellated by a range of ideological systems which means that often people inhabit quite conflicting subjectivities which are experienced differently in different contexts. For example, given the hegemonic construction of Western motherhood as kind, gentle and warm, a pregnant woman wearing 'dirty' jeans, a ripped T-shirt, and studs in her tongue and eyebrows would more than likely be interpellated by other pregnant women at a birth class as a 'bad' subject even though she may have a deep investment in being a 'good mother'. As Probyn (2003) argues, subjectivity is produced, and the site and space of its production is vitally important.

In recent years geographers have begun to recognize that this notion of shaping and being shaped by social space also relates to fashion and the ways people clothe themselves. In the early 1990s geographers interested in fashion and clothing tended to focus on production and consumption processes (e.g. Crewe and Davenport 1992; Crewe and Forster 1993) but more recently the focus has shifted on to relations between fashion, clothing, subjectivity, and spatiality (e.g. Crewe and Lowe 1996; Dwyer 1999; Gregson *et al.* 2001).

Despite the growth in interest in fashion and clothing, academics have had very little to say about dressing and adorning the *pregnant* body. Even books that focus on gender, femininity, and women usually contain no mention of pregnant bodies or of maternity wear. Academics, it seems, have assumed pregnant women to be 'outside' of fashion. Fortunately, however, there are some exceptions to this general exclusion.

Rebecca Bailey (1992) presents an interesting account of the relationship between maternity clothing, pregnant women's occupation of private and public spaces, and advice offered to pregnant women in the United States, 1850–1990. One of the most interesting points raised by Bailey for the purposes of this chapter is that 'there was no public sanction of appropriate maternity dress outside of a private dwelling during daylight until 1911, when the *New York Herald* agreed to run an advertisement for the previously unmentionable goods [maternity apparel] of the Lane Bryant Company' (Bailey 1992: 255). The advertisement stated that it was no longer the fashion for expectant mothers to stay secluded in private space. The Lane Bryant shop was mobbed by excited pregnant women who, by the end of the day, had stripped the street-wear garment rack empty (Bailey 1992).

Gregson and Beale (2004) focus on the maternity wear and clothing practices of contemporary pregnant women in the United Kingdom. They carried out sequential interviews with 20 pregnant women and some new mothers living in Manchester, Bradford and Northampton. Gregson and Beale (2004: 689) draw on these data to argue that it is important that accounts of clothing, performance and identity do not simply 'restrict themselves to what might be termed the "present-ness" of wearing and dress' or the 'visible, worn assemblages of clothing and their meanings' because this overlooks the complexity that surrounds wearing. It is about the clothes in the wardrobe that have not been thrown out even though they are no longer worn (Banim and Guy 2001), it is about passing on clothes to others (Beale 2003), it is about the anxieties of wearing (Clarke and Miller 2002), and it is about the anticipation of wearing (Green 2001).

My research indicated that there are similarities between pregnant women in the United Kingdom and New Zealand but before commenting further on the experiences of a group of pregnant women in New Zealand, I discuss the methodological process undertaken to conduct the research.

DATA COLLECTION AND ANALYSES

The same data are used in this chapter as in Chapter 2 (also see the Appendix). In addition to the 200 pages of transcripts, however, I also compiled a folder of maternity wear brochures, advertisements for maternity wear, newspaper clippings, magazine photographs and stories (mainly from *NZ Woman's Weekly*, *NZ Woman's Day* and *New Idea*[1]) and information from Internet sites selling maternity wear/fashion. In relation to the visual images of pregnant women I noted the similarities and oppositions between the images paying attention to whom the women were with and their pose, facial expressions, clothes, jewelry, hairstyles, and make-up. My analysis was guided by Gillian Rose's (2001) discussion of discourse analysis of visual images because the relationship between 'visual images, verbal texts, institutions and social practices' (Rose 2001: 140) was my central concern.

Finally, in the course of my everyday interactions, over a period of several months while conducting interviews, I observed pregnant women in public and semi-public places such as at shopping malls, supermarkets, and swimming pools. Most participants were interviewed in their homes but I was also interested in how pregnant women perform their selves in public so wanted to observe what they were wearing, where they were shopping, who they were with, and what activities they were engaging in. At times, watching pregnant women stirred memories of my own pregnancies. In a sense, I relived the bodily sensations of when I was pregnant and reflected on what I had worn and how I had felt about my body and my clothing. In this chapter I do not discuss these data explicitly but they provided a useful context for the research.

FROM MATERNITY WEAR TO MATERNITY FASHION: OR FROM SMOCK TOPS TO CROP TOPS

In the 1950s and 1960s smock tops with ample shirring or gathering were standard attire for pregnant women in New Zealand. In the 1970s and 1980s, and even in the early 1990s, 'tent dresses' and baggy T-shirts were 'popular'. Such garments aimed at veiling, or disguising, the pregnant body (see Wardell Morrone 1984). Pregnancy was not a time for *haute couture* or for adopting radical looks from the street. It was a time to dress conservatively, comfortably, and modestly. Maternity wear was aimed at disguising the pregnant form. Women's focus was expected to be on interiority rather than exteriority. The social expectation was that women's 'essential beauty' would radiate from within during pregnancy. Women were expected to collect goods and shop for the coming baby but not to spend huge amounts of time and money on adorning their own bodies.

Advertising and retailing industries, however, are always looking for new markets and while in the 1980s marketers began selling the idea of beauty, fitness, and health to young men (Miller *et al.* 1998; Mort 1996), in the 1990s marketers began seeing pregnant women as offering commercial possibilities. Midge, the pregnant doll, is an interesting example of this commodification of pregnancy in that she is a product available for purchase in many Western countries (Mattell is likely targeting white, middle-class pregnant women, mothers, and daughters) but additionally the doll's embodiment reflects that of a white, middle-class, heterosexual, fashion conscious, pregnant woman. In this way Midge represents a sort of 'double-commodification' of pregnancy.

In the 1990s a new term entered the 'everyday' English language — 'pregnancy chic'. Just prior to the 1990s, Lynn Sutherland with Audrey Brooks (1989) published a 'popular' book entitled *Pregnant and Chic* that shows readers 'how to keep looking your best — even with a 44-inch waist line' (front cover). In 1998 Cherie Serota and Jody Kozlow Gardner published their 'popular' 'fashion survival guide' entitled *Pregnancy Chic*. In April 2001 doctors and nurses took to the cat walk at the Royal Opera House in Covent Garden, London, for the Charity Fashion Show for mothers and mothers-to-be. The website describing the Show (Maternity Fashion Show, 2003, available <htpp://www.uclh.org/news/bumps0401.shtml>, accessed 4 November 2003) claims: 'There will be no place for flat tummies as this show is aimed at fashion conscious pregnant women ... expectant mothers can dress to impress.'

Currently there are numerous Internet sites selling maternity garments with many offering clothing for the 'fashion inspired'. For example, Maternity Clothing Fashions in Ohio state: 'Just because you're pregnant doesn't mean your stylish days are over ... we offer the look of today's runway fashions ... Big shirts and tent-like dresses are maternity clothing history' (Maternity Clothing Fashions, 2006, available <htpp://www.maternity-clothing-fashions.com>, accessed 2 November 2006). Maternity lingerie (not just nursing bras and underwear) is also available, including the 'maternity thong' (Maternity Thong, 2003, available <http://www.new-port-news.com/shop/Product.asp>, accessed 20 November 2003). The LSR maternity collection also offers sleepwear in a range of 'looks' — peaches and cream, flirty, classic, shagadellic, alluring, and romantic (LSR Maternity Collection, 2003, available <http://www.lsrmaternity.com>, accessed 20 November 2003). The website boasts creating designs for a number of Hollywood celebrities.

In 1992 a lingerie shop called Hot Gossip in Hamilton, New Zealand, featured slim, white, youthful mannequins clad in red and black suspender belts and lacy bras in their window display. One of the respondents entered the shop to buy some of this 'sexy' lingerie (rather than a feeding or maternity bra) and received a 'frosty response' from a shop assistant. In February 2003 Hot Gossip featured a pregnant mannequin in their window display (see Figure 3.2). The mannequin (who, unlike the other mannequins in the display, was just a torso) was dressed in a black bra and black full brief. There were no pregnant mannequins in window displays in Hamilton ten years ago.

Stories and photographs of pregnant celebrities in Western women's magazines have played a significant role in turning maternity wear into maternity fashion (Tyler 2001). This trend began in 1991 when Hollywood actor Demi Moore appeared pregnant and naked on the front cover of the glossy magazine *Vanity Fair*. Prior to Moore's appearance a strong cultural

Figure 3.2 Pregnant mannequin in 'Hot Gossip' bra shop. *Source*: Photograph by Amanda Banks.

taboo had regulated popular media representations of pregnant women (Tyler 2001).

Since the Demi Moore shots in 1991 numerous other actors, singers, and models have followed suit and posed pregnant for photographs in women's, fashion and 'lifestyle' magazines. Jones (1999: 35 cited in Tyler 2001: 69) states 'posing for photoshoots while pregnant has become an almost compulsory celebrity rite of passage.' For example, the sample of approximately ten issues of the three popular magazines I examined in 2002 — *NZ Woman's Weekly*, *NZ Woman's Day*, and *New Idea* — contained an exposé on Sarah Jessica Parker's (star of the television series *Sex and the City*) 'pregnancy wardrobe', photographs of a 'glowing' Cate Blanchett 'out and about with husband Andrew', a story about and photograph of New Zealand TV presenter Jeanette Thomas expecting her first child, and a story about Jools, wife of the British 'Naked Chef' Jamie Oliver, who was expecting her first child.[2] Joan Smith (2003: 2), a journalist for the New Zealand *Weekend Herald*, reports, in 2002 'the image of the pregnant woman underwent a complete makeover. If you were having a baby last year, you could expect to look like a supermodel for most of the nine months.' In short, societal expectations of how women 'do pregnancy' in New Zealand are changing and this is both reflected and reinforced through clothing practices often represented in popular media. Over the past decade maternity wear has become maternity fashion — a term that less than a generation ago would have been considered an oxymoron.

In the section that follows it is argued that for the pregnant women in this research maternity fashion has allowed for new engagements with a range of complex and often contradictory gendered subject positions. Style, fashion and dress have allowed for a shift from a homogenous image of the pregnant woman to more diverse images of pregnant women. Pregnant women now appear to have more opportunities to experiment with a range of subject positions using clothing to conceal and reveal particular bits of their bodies as they see fit and to inhabit a greater range of spaces. I move now to discuss four of the main 'looks' and subjectivities that participants in this study 'tried on' as they moved towards becoming mothers for the first time.

TRYING ON DIFFERENT LOOKS AND DIFFERENT SUBJECTIVITIES

Although stylish maternity clothes now figure much more frequently in women's magazines, on the Internet, in catalogues, and in specialist stores, none of the participants in this study had invested, or planned to invest, 'heavily' in maternity wear, even those on six figure incomes, such as Angelina who works in the media. Angelina is highly aware that it is possible to utilize consumer culture to create a maternal self that appears stylish, wealthy and powerful but preferred to spend more modestly on maternity wear. Before becoming pregnant she invested a great deal of time and money in clothes. Now, however, she figures: 'It's a bit of a waste buying clothes that you are only going to wear for a few months.' Sandra, an insurance claims assessor whose household income was close on NZ$200,000 (US$125,850), also concurs: 'I didn't want to spend a fortune on clothes.'

The trend is also noted by Gregson and Beale (2004) in the context of pregnant women in the United Kingdom although they comment that for at least one of their participants, who purchased only five maternity garments, the motivations were not exclusively about thrift or the elevation of baby things over things for pregnancy but about an 'acute ambivalence' towards pregnant corporeality.

Participants in Hamilton discussed feeling uncomfortable with their new growing bodies and mentioned having 'lumpy thighs', a 'big bum', 'big boobs', stretch marks, and a stomach that was 'funny looking' and not tanned. Discussion, however, also focused on the lack of appeal of much of the maternity wear in Hamilton rather than just on negative body image. Carla, a university lecturer, notes: 'I do wish there was better looking maternity wear'. She searched the Web for maternity clothing but 'most of the sites are in the United States and they won't ship to New Zealand.' Trish, a teacher, thinks maternity wear in Hamilton is 'quite matronly,' or alternatively, it's 'like grown up baby clothes — dungarees and stuff like that.' Trish had recently been in London and bought 'trendy preggy clothes' in maternity ranges in High Street stores such Dorothy Perkins but on returning to New Zealand found the maternity wear to be 'quite disappointing.' In contrast, Nia, a natural therapist, comments there are 'heaps of beautiful looking maternity clothes but I can't afford them.'

At least half of the participants considered stylized or designer label maternity fashion to be overpriced and preferred instead to save their money, buy from cheaper maternity ranges offered in some department stores, or spend the money on the imminent arrival. Gregson and Beale (2004: 689) say that discourses of 'thrift, sacrifice and making do' are common amongst pregnant women. It is perhaps not surprising therefore, that in addition to limiting purchases of maternity garments, and/or purchasing reasonably inexpensive garments, pregnant women also exchange garments amongst themselves as a way of being thrifty and making do.

Eva, aged 31 who works in a dairy (small local grocery store) borrowed a few 'nice, expensive clothes' from friends. Like Eva, Jennifer, a public sector scientist, 'didn't really want to go out and buy a whole new wardrobe' and found that friends with toddlers were 'sympathetic' and 'supportive' not just offering books and someone to talk to but also maternity wear. Dimi mentions that friends who have children have 'been so generous. I've gotten heaps of clothes already — maternity clothes and baby clothes.'

Gregson and Beale (2004) argue that the circulation of maternity wear is itself a marker of pregnancy and motherhood regardless of whether this clothing is worn or not. It is 'an embodied divestment/recovery ritual' (Gregson and Beale 2004: 690). Gregson and Beale note that although women sometimes keep clothes they have not worn for a long time and are unlikely to wear again (e.g. a wedding dress) they seldom keep maternity wear in their wardrobes.

But a picture of becoming mothers as wearing cheap or hand-around clothes, making do, and not being overly worried about appearance needs to be complicated. A somewhat contradictory, recurrent theme in this research was that despite this trend participants indicated they are very aware of fashion and all but two articulated a desire to wear trendy, fitting and sometimes revealing clothes during pregnancy. Interestingly, the pregnant body, not long ago coded as abject (see Longhurst 2000 and 2001a), was described by Olivia as having become 'like a fashion accessory — it's something to be proud of'.

One of the clothing fashions in Hamilton in the summer 2002–03 (when most of the interviews were conducted) being adopted mainly by women under 30 years was low-riding 'hipster' pants or jeans worn with short, fitting, stretch or chiffon tops. Skirts too were popular, and often worn with cropped tops and bare midriffs. It was not only 'slim' women adopting this 'look' but also some larger young women. This is a manner of dress that can be adopted by pregnant women because pants and skirts with a low waist can sit beneath a pregnant stomach, while tops made of stretch fabrics can fit across the stomach. This look is represented in many women's fashion and lifestyle magazines.

It is impossible to ascertain exactly what effect these magazines have on their readership but 28 year old Trish thinks 'you are influenced by people in the media, like, I mean all those pop stars ... wearing tight things when they're pregnant and it's like, fashion, basically.' Olivia says the photograph of Demi Moore on the cover of *Vanity Fair* 'was not disgusting but more like "Wow!" and it sticks in everybody's mind — even my husband's!'

The participants were also influenced by what other pregnant women whom they knew or had seen in public spaces were wearing. Participants commented on what they thought looked 'nice', 'cheap', 'respectable', 'stylish' and so on. Trish, for example, said sometimes she imagined herself trying on the different 'looks' she had seen. Sometimes she was motivated to go and buy a similar garment. Sometimes participants thought particular styles (such as revealing the lower section of a naked heavily pregnant stomach) looked 'good' on others but they would not adopt that look themselves.

Of the 19 participants in this study, only two (Jennifer and Denise) prefer wearing loose garments mainly because they are comfortable. Another three participants (Mihi, Nia and Sarah) said they would like to wear fitting or tight garments but feel too self-conscious about what others might think. Two of these five women who wear loose garments (Mihi and Nia) report not going out in public nearly as much during pregnancy as they used to — Mihi because she feels there is a social stigma attached to being a sole-teenage mother and Nia because she feels tired and self-conscious about her body and clothes. The remaining 14 at various times and places wear fitting or tight garments that expose their stomachs. All seem comfortable in a variety of public spaces. Participants disagree as to whether loose fitting garments or tighter garments that expose more flesh are coolest in the heat of summer (which is when the interviews were conducted). The participants wear a range of colours, although black 'bottoms' (skirts, trousers, three-quarter pants, shorts) is a popular choice on account of black being a 'slimming colour' and easy to mix and match with variety of tops. There is also a tendency amongst the women to want to wear the same kind of clothes as before they were pregnant. Jennifer comments 'if you didn't wear crop tops before you were pregnant you are probably aren't gonna start now.'

Having outlined a few general trends in the data, it is now useful to focus on some specific comments made by participants. Kate, a 30 year old primary school teacher who was entering the final month of her pregnancy, comments: 'I think I look and feel better while wearing the tighter things'. Eva who was 22 weeks pregnant at the time of interview reiterates Kate's sentiments: 'I'd much rather show it off ... I've got a black one of these [she points to the top she is wearing] ... and it's more fitting than this. It's half the size and you can tell that I've got a *puku* [Māori word for stomach].

Carla wears a two piece bathing costume (bikini): 'I refuse to get in a one piece suit. It seems so stupid, you know, it's like I own two piece suits, they're much more functional.' When Eva is in public space, although not at work, she likes to 'show it [her enlarged breasts and stomach] off' and wears fitting tops, often black and/or with mesh. She comments: 'My husband now tells me that instead of having one handful, I've got two in one! ... It still turns him on and he still tells me I look beautiful'.

In contrast 26 year old Nia comments that since becoming pregnant her husband has been 'turned off' by her (this supports Bailey's 2001 study of 30 middle-class, white, first-time mothers who claimed that pregnancy seemed to erase part of their sexuality). Nia stopped going out as much. She explains: 'I've felt so ugly ... and I feel kind of disappointed at myself for not living up to that expectation of being a beautiful pregnant woman.'

It is not easy to live up to the beautiful and fashionable pregnant icons in the glossy magazines. 'Everyday' women do not always look fantastic in their clothes. Clothes get stained, crumpled and faded (see Banim *et al.* 2001). Also, many 'everyday' pregnant women worry about appearing fat.

Half of the participants (all of whom I would describe as being of 'average' build) explained that their desire to wear fitting clothing in the early stages of pregnancy was in part motivated by a desire to look fashionable but also by a fear that their bodies would be read by strangers as fat or as carrying 'extra weight' (a concern that seems to be shared by some women in the United Kingdom — see Earle 2003). Kate explains that in the first trimester:

> You can tell people look at you, and are wondering, especially at that time when it's sort of in the in-between stage. You can almost see them weighing up 'is she pregnant or is that just a bit of extra weight?' [laughter].

Braziel and LeBesco (2001: 135) argue that in Western societies 'fat bodies, and fat women's bodies in particular, are represented as a kind of abject'. Kate did not want her body to be read as abject. Nia reduced the risk of people thinking she was fat by carrying a bag from a baby retail outlet so that people would know that she was pregnant rather than fat.

At least half of the participants (like those in Earle's 2003 study of 19 pregnant women's experiences of body size, shape, and weight during pregnancy) expressed relief when their stomachs were finally large enough, and firm enough, so that their pregnant status could be easily recognized. Trish explains:

> It's the first time my belly's been hard for a long time! It's generally a lot flabbier ... And, yeah, so I'm quite happy going out in public, I'm not feeling all blobby and fat.

I would not describe any of the participants in this study as 'fat', or to use the more clinical term, 'obese'. Rosemary Wiles (1994), however, interviewed 37 'fat' women in the United Kingdom during pregnancy. Forty one percent felt better about their weight during pregnancy probably because there appears to be a greater acceptability of fatness during pregnancy.

Perhaps it is not surprising, therefore, that in contrast to those participants who felt concern about appearing fat, four of the participants felt that pregnancy freed them from the beauty, fitness and eating regimes that so many women often feel subjected to (see Bordo 1993; Chapkiss 1986; Cline 1990). Angelina explains:

> I've always been really concerned about my weight and body issues, things like that. You know, going to the gym and making sure you eat right and all that sort of thing. Whereas [now], it feels like, I've been given the keys for the freedom of the city [laughter].

One space that participants appear to be inhabiting much more freely and frequently than their counterparts a decade ago is the workplace. Sixteen of the 19 participants were still in paid employment at the time of interview. Of these 16 at least 10 needed to dress 'formally' for work. This was especially the case for those participants involved in business or in professional occupations.

As was suggested in the previous chapter Carla was keen to present a 'professional image' when lecturing (see Green 2001 on women professors using clothes to signal authority). Sandra, who assesses claims for an insurance company, comments: 'we're supposed to be business-like. They probably wouldn't have worried too much if I'd worn more casual clothes but I had it in my mind that I wanted to maintain a certain standard of dress.' Sandra spent approximately NZ$600 (US$378) on maternity wear, mainly 'nice clothes to wear at work.'

One way of extending the time that it is possible to wear one's 'non-pregnant' clothes for longer (which is especially useful for formal and business attire which tends to be more costly than casual attire) is to use a 'belly belt'. The belly belt is an insert which is available in a range of colours. It is extendable and fastens with a button and loop of cord or elastic into the waistband of jeans, skirts, trousers and shorts. Several participants including Carla said they had purchased a 'belly belt' which is available in most stores selling maternity wear. Wearing a belly belt can delay the subject being marked as pregnant because she is able to wear her 'normal' clothes for longer.

Wearing 'normal' clothes at work can be useful because, as Davis-Floyd (1994) notes, femininity, and more specifically pregnancy, 'violates' the separation of the personal and professional life; pregnancy drags the body and the personal into the workplace (see Bailey 2001). Wearing professional attire is one strategy that pregnant women adopt in order to ward off constructions of them as unprofessional, irrational, emotional, and unfit to keep working (Longhurst 2001b).

Clothes that have long been used to signal women's power and authority at work such as well pressed straight skirts, trousers and jackets, and shoes with a heel (but not too high as to be 'sexy') can become increasingly tight and difficult to wear for pregnant women and this can lead to many ambivalences, dilemmas, and worries about their appearances at work. Many women want to remain 'credible workers' but when their 'credible clothes' no longer fit they must acquire new clothes and forge new subjectivities.

CONCLUSION

A decade ago pregnant women in New Zealand and in many other Western contexts could legitimately wear baggy and unfashionable garments. They could get by with 'hand-me-downs' or 'hand-me-arounds' and escape body and beauty regimes. This is not the case for many pregnant women living in the West today. By privileging the creation of looks, fashion and image in the production of subjectivity, pregnant women are increasingly being expected to reinforce the norms of a consumer society that offers the possibilities of a newly created self through consumption and the products of the fashion industry.

Increasingly pregnant women are being expected to look attractive, well groomed, fashionable, and 'sexy' and for some women this has added to what are already heavy expectations on pregnant women. Today not only are pregnant women expected to eat healthy and nutritious food, give up alcohol, drugs and smoking, and exercise gently but regularly, they also shoulder expectations (their own and others) to look like the glamorous celebrities featured in the latest glossy magazines. Rather than opening up possibilities, 'pregnancy chic' represents for some women a new set of pressures to perform the self in yet another tightly prescribed manner.

However, representations of 'pregnancy chic' have also acted as a catalyst for some becoming mothers to try on new styles of clothing and new subjectivities. This has led to a new visibility of pregnant women in public space such as workplaces, cafes, restaurants, and sporting arenas. Some pregnant women have thrown away the idea that the pregnant body must be hidden and have instead adopted a fashionable and sometimes revealing style as a way of resisting constructions of the pregnant subject as modest, 'respectable', domestic, private and associated with nature.

Clothes matter, however, not just for pregnant women. Clothes are used to construct different images, at different times, and in different spatial contexts. Fashion and clothing are cultural constructions of embodied subjectivities — subjectivities which are intimately bound to spatialities. Lifestyle, social class, gender, sexuality, and ethnicity become expressed through the production, consumption, wearing, storage, and circulation of clothes in particular spaces. I have examined women's pregnant bodies in New Zealand but the meaning and use of clothing and fashion could also be examined in relation to other bodies such as athletic bodies, fat bodies, 'poor' bodies, academics' bodies, and disabled bodies. In short, until recently the idea that fashion and clothing matter has not been widely considered by geographers, and yet this area offers huge potential for understanding more fully the complex relationships between the self, the social, and the spatial.

4 Pregnant and Disabled

'Body Troubles'?

This chapter continues the theme of pregnancy but provides a different window on maternities, gender, bodies, and space. It examines the experiences of pregnant women with physical disabilities including, but not only, conditions such as arthritis (which can go into remission when women are pregnant), amputations, cerebral palsy, degenerative disk disease, fibromyalgia, congenital conditions, lupus, multiple sclerosis, neuromuscular disorders, spina bifida, and spinal cord injuries. Women with physical disabilities are often as fertile as women without disabilities but disabled women who decide to get pregnant sometimes encounter opposition, discrimination and a great deal of skepticism (Prilleltensky 2003). Some people may be surprised that a disabled woman would contemplate having sex, let alone have a baby.[1]

The basic argument mounted in this chapter is that pregnant women with physical disabilities inhabit some of the same emotional, discursive and material spaces as non-disabled pregnant women. They also, however, inhabit some 'different' spaces. The chapter is written from the perspective of a non-disabled woman. Not having walked in the shoes of a disabled woman my understanding is likely to be limited but as researchers it is not always imperative to first live what we study. I did not want to exclude the experiences of disabled pregnant women from the book because it seems that their experiences provide useful insights into the array of different constructions of motherhood. Whereas countless women in many parts of the world experience social pressure to *have* children, disabled women often find they are under pressure *not* to have children (Rogers 2006: 1).

One of the most useful sources of information drawn upon extensively in this chapter is Judith Rogers' (2006) *The Disabled Woman's Guide to Pregnancy and Birth*. Rogers has been disabled since birth. She argues that in recent years there has been greater acceptance of disabled people's sexuality and sexual needs. 'The next step has to be validation of the reproductive capacity of disabled women' (Rogers 2006: xi). When Rogers became pregnant she found there was hardly any literature that addressed her particular needs. There is a vast literature on pregnancy, labour, delivery and child rearing in general but little of this literature is given over to addressing the specific needs of the disabled.

Rogers collected together academic work and stories from disabled pregnant women in an attempt to further understand their lived experiences. In order to write this chapter I read Rogers' book (it is a weighty 500 pages) and a range of other mainly academic texts (although work in this area appears to be rather limited) on the experiences of disabled pregnant women and mothers, paying careful attention to issues of spatiality. I looked for information regarding the emotional, discursive and material spaces inhabited by physically disabled pregnant women.

The chapter is situated theoretically within the field of feminist disability studies, which

> is more than research and scholarship about women with disabilities, just as feminist scholarship extends beyond women to critically analyse the entire gender system. Like feminist studies itself, feminist disability studies is academic cultural work with a sharp political edge and vigorous critical punch. (Garland-Thomson 2005: 1557)

Feminist geographical work on disability also informs this chapter. I aim to build on the work of other feminist, social, cultural and health geographers such as Vera Chouinard and Ali Grant (1995), Ruth Butler and Hester Parr (1999), Joyce Davidson (2001), Isabel Dyck, Nancy Davis Lewis and Sara McLafferty (2001), and Pamela Moss and Isabel Dyck (2002) who have examined women's experiences of illness, impairment and disability. Disabled pregnant women often face 'body troubles' (Dyck 1999) because dominant discourses dictate that their bodies — bodies that are considered unreliable, unsightly and sick — should not reproduce. Over the past few years a literature on disabled parents (and their children who often take care of them) has begun to emerge (see Grue and Tafjord Lærum 2002; McKeever *et al.* 2003; Prilleltensky 2003; Thomas 1997) but still little has been published on disabled pregnant women. This chapter will add to these debates by exploring the 'body troubles' but also the joys and pleasures experienced by physically disabled pregnant women.

Numerous disability activists and scholars have over the past decade questioned the assumption that disability is a lack, an excess, or a failing of some sort (see Gleeson 1999; also see Chapter 8 this volume on constructions of subjects as lacking). Instead disability has come to be understood in broadly social rather than medical terms (Irmie 1996; Oliver 1990). Geographers have stressed the importance of understanding not only the intersections between impairment and sociality, but also space. Dyck (1995: 308) explains:

> Close attention to the body in material context provides the potential for exploring the involvement of dominant discourses and power rela-

> tions in the social construction of ideas about the body and identities, including that of the 'disabled body' and the implications of the experience of place.

A complex relationship exists between place, maternity and disability and much work remains to be done in this area. My contribution is limited to examining concepts of body and space in an attempt to better understand the daily lives of a few women who are pregnant and physically disabled.

Rogers' (2006) book has been invaluable in helping with this task. She has gathered together 'the stories of 90 women with disabilities who chose to have children and who were willing to share their experiences of pregnancy through the postpartum period' (Rogers 2006: 1). Rogers also provides an appendix that contains these 90 women's detailed pregnancy histories. Issues of space and place, however, are not mentioned in all these histories.

In short, the chapter attempts to offer insights on the embodied experiences of some pregnant physically disabled women and the emotional, discursive and material spaces that they shape and are shaped by. It begins with a discussion of the emotional spaces and how disabled pregnant women often feel both affirmed and disapproved of for becoming pregnant. Second, the chapter examines the ways in which the eugenics movement created an oppressive discursive space for disabled women and their offspring that constructed them as lacking. The sentiments of eugenics have not disappeared entirely today even if the science has been discredited. However, disabled women have resisted the societal quest for bodily perfection and instead taken up their rights to become mothers. Finally, the chapter looks at some of the difficulties faced by disabled pregnant women and mothers in relation to mobility and the disabling effects of some material environments. Negotiating the material spaces often associated with motherhood such as homes, clinics, birthing wards, and shopping areas can sometimes be a challenge for disabled pregnant women and mothers.

EMOTIONAL GEOGRAPHIES: NEGOTIATING NEW SUBJECTIVITIES

Over the past few years geographers have begun to pay more attention to 'emotional geographies' (see Davidson, Bondi and Smith 2005, and the special issue on emotional geographies in *Social and Cultural Geography* 2004, Vol. 5, No. 4). In the editorial to this special issue of *Social and Cultural Geography* Joyce Davidson and Christine Milligan (2004: 523–524) claim:

> Our emotional relations and interactions weave through and help form the fabric of our unique personal geographies. We live in worlds of

> pain or of pleasure; emotional environs that we sense can expand or contract in response to our experience of events ...

This is especially evident in some of the geographical research on disability. For example, Ruth Butler and Hester Parr's edited collection *Mind and Body Spaces* (1999) contains a number of contributions that focus on the emotional highs and lows that go along with continual efforts to negotiate (dis)abling social and material spaces. Contributions such as Gill Valentine's (1999) on 'what it means to be a man,' Isabel Dyck's (1999) on 'women, the workplace and negotiations of a disabled identity,' and Hester Parr's (1999) on 'different geographies of mental health' are testimony to the emotional geographies of those who live with illness, impairment and disability.

The emotional geographies of pregnant physically disabled women illustrate that at times these women feel excluded from the mainstream on account of their disability. At other times they feel accepted as part of a new community of mothers. Examining pregnant physically disabled women's pains and pleasures as experienced across a range of sites and scales reveals a complex geography of continual (re)negotiation with self, loved ones, family, friends, co-workers, and strangers. It quickly became apparent when examining the narratives of pregnant physically disabled women that their experiences and emotional geographies vary enormously — a point that is borne out in the narratives that follow.

Deborah Kent is blind. From a young age she had always dreamed of becoming a mother but her blindness marked her as 'different'. She felt she had no role models to follow. When Kent finally became pregnant she was thrilled.

> I loved being pregnant. Unlike those Victorian ladies who went into confinement, disappearing behind whispers and closed doors the moment their 'delicate condition' became apparent, I wanted the world to take notice of my bulging belly. I seized every opportunity to walk the streets, to ride the city buses, to present myself in public. I revelled in the anticipation of motherhood, and somewhere too, I felt an exhilarating sense of defiance. (Kent 2002: 81)

When Kent became pregnant she felt healthy and happy, and received affirmations from supportive parents and friends for fulfilling the role of mother. However, she also suffered prejudice, self-doubt and anxiety about her ability to be mother, and a fear of others considering her to be an inadequate parent on account of the fact she is blind.

Kent's narrative relays some of the highs and lows, the joys and sorrows, of embarking upon motherhood. Her story of becoming pregnant is useful because she identifies a number of issues that arose in many disabled women's accounts of pregnancy. Kent reflects:

> No one ever told me point-blank that I couldn't have children. Nobody had to say it in words. From childhood on, I heard the message in a subtext of denials and omissions. Nearly all of the women I knew were mothers, but not one of those mothers was blind. It was always a sighted mom who pushed the stroller, ran the Scout troop, or called to her kids over the backyard fences. Throngs of sighted mothers filled the auditorium for music programs at school. I knew only two adult blind women, and both of them were childless. In my young mind, this sample translated to a global truth. Sighted women were mothers; blind women were not. (Kent 2002: 81)

Although words of prohibition were never uttered, Kent from an early age felt excluded from the spaces of motherhood. Fortunately this did not deter her and at age 34 she became pregnant and took pleasure in joining a new community of mothers. Kent (2002: 83) says:

> No longer was my blindness the primary focus of attention when I met someone for the first time. Now, instead of my disability, people talked about pregnancy and parenting. These things they understood, valued, and celebrated. My blindness receded into the background. I was no longer an outsider; I was one of the initiate at last. (Kent 2002: 83)

Others with physical disabilities report similar experiences. Grue and Tafjord Lærum (2002) claim that a theme that emerged from their interviews with 30 mothers with physical disabilities in Norway was that disabled women become accustomed to their bodies receiving a great deal of medical attention, focusing on what is wrong with them, and what might enable their bodies to function better. The mothers in their study commented that when they became pregnant their bodies were made into something of 'great value, something capable of producing new life' (Grue and Tafjord Lærum 2002: 676). These mothers also reported that for those who had been physically disabled since childhood, becoming a mother meant that 'for the first time in their life they experienced being looked upon and being met by other people as an adult person and not just as a disabled person. Becoming a mother made them achieve womanhood' (Grue and Tafjord Lærum 2002: 676).[2]

Ora Prilleltensky (2003) conducted interviews with 13 women and four focus groups with a total of 25 physically disabled Canadian women. Eight of the 13 interviewees were mothers with children of different ages. The other five women were without children. Similarly, two of the four focus groups were attended by mothers, the other two by women without children. Most of the participants had mobility/limb impairments and used a wheelchair. Prilleltensky is herself a mother with a physical disability (muscular dystrophy). Amongst other things, participants described reactions to their pregnancy — their own reactions, and the reactions of others.

Some of Prilleltensky's participants were delighted to discover they were pregnant. 'One woman who was told for years by her family that she would not become a mother, described a sense of elation that stayed with her throughout her pregnancy' (Prilleltensky 2003: 26). She said that when she learned that she could become a mother and the gynecologist said to her 'you know, you're normal' she started to cry (tears of joy) because she'd always been told that she wasn't 'normal'. Another participant told Prilleltensky (2003: 26) 'I do conceive really easily ... that's the one part of my body that seems to work ... my eyes and my uterus ... I always figure, everything else is going to disappoint me, nothing else is completely reliable, but these two work well ... it feels great.'

One of the most poignant articles I read on the topic of disability and pregnancy was Ann Neville-Jan's (2004) autoethnography of her quest for a child. Neville-Jan is an academic with spina bifida who lives in the United States. She writes 'For me, giving birth to a child became the ultimate life project that would demonstrate I could achieve the same goals as normal people' (Neville-Jan 2004: 117). Neville-Jan's first pregnancy came to a tragic end when, due to complications associated with amniocentesis, her baby daughter was born at 20 weeks gestation. She did, however, become pregnant a second time and gave birth to a healthy son. At a similar time, Neville-Jan and her husband adopted a daughter from China. What makes Neville-Jan's article so powerful is that she writes as both a researcher and research participant noting that she 'became aware that a gap of representation existed in the scholarly storyline of pain, the voice of the person in pain' (Neville-Jan 2004: 114). Neville-Jan speaks frankly about her pain and her sexuality but also about her success in her quest for a child.

Many of the participants in Rogers' (2006) study also talked about their successes in their quest for a child or children and about feeling affirmed when they became pregnant. She reports:

> Julie [said, I] 'loved being pregnant because for once my body worked right.' Arlene said, 'Having a baby made me less handicapped because I was able to fulfill one of the female roles in society and I was really rewarded for it.' Heather felt that she was 'proving I was as independent and as self-reliant as anyone.' Noelle 'loved being associated with the non-disabled population. It was my only chance,' adding that the shared interests and concerns of parenthood offered a basis for friendships with non-disabled people. (Rogers 2006: 37)

Grue and Tafjord Lærum's (2002: 676) research reveals that for many women, depending on when they had become physically impaired, having a child meant either 'capturing' a gender, or 'recapturing' a lost gender. Some of the participants reported having found it difficult during adolescence and early adulthood to stage convincing gendered performances as women. One particular participant in Grue and Tafjord Lærum's (2002) study, Kari, puts it like this:

> I instantly felt more like a woman when my son was born. During my whole life I had felt like a person without a gender. The process I went through during pregnancy and afterwards made me more conscious of my identity and my worth as a human being. My child made it possible for me to say both to myself and in relation to other women — see, I am the same as you. (Grue and Tafjord Lærum 2002: 676)

Despite these positive reports from physically disabled women who felt a strong sense of affirmation — a shift in identity from a disabled person to a mother — their decisions to become mothers were in some instances extremely complex and difficult to make. Rogers (2006: 33) points out that most books that aim to help people decide whether or not to become parents do not address the concerns of women with disabilities. Also, social disapproval and disabled women's own fears of becoming a parent tend to work against many disabled women in their decision to have children. From the outset then, pregnancy, birth and motherhood can be difficult terrain to embark upon for disabled women. There is a long history of disabled women feeling under pressure to, and even worse, being forced to, be sterilized or abort their unborn babies.[3]

DISCURSIVE GEOGRAPHIES: POWER AND RESISTANCE

Prilleltensky's (2003: 22) research reveals that not all physically disabled women feel affirmed when they embark upon motherhood. Some face opposition, skepticism and even hostility.

> As a group, women with disabilities have been traditionally discouraged or even denied the opportunity to bear and rear children. Whereas other women experienced societal pressure to have children as oppressive, women with disabilities were perceived as child-like, dependent and asexual and were excluded from fulfilling traditional female roles. (Prilleltensky 2003: 22)

A number of authors have now documented stories of disabled women feeling incredible pressure to refrain from reproducing, or to have abortions, hysterectomies, or tubal ligations (e.g. Gill 1996; Rogers 2006). Prilleltensky (2003) tells the story of a participant in her research who was single and unexpectedly found herself pregnant. The woman was initially upset but soon formed a strong connection with her unborn baby and decided not to terminate the pregnancy even though many tried to persuade her to take this path.

> A number of people said that to me ... I was having nurses coming in and out because I was having my blood taken ... it would be like ... say you have a cough, 'maybe you should have cough medicine' ... that's how casually they were saying it, 'maybe you should have an abortion, maybe it would be for the best' ... they thought his life wasn't important. (participant quoted in Prilleltensky 2003: 27)

Prilleltensky (2003: 27) argues, and I agree wholeheartedly, that the demands put on this single disabled woman to abort her possibly disabled fetus need be understood within the context of eugenic ideology.

Eugenics or 'racial hygiene' — the term used in Nazi Germany — questions the value of caring for the 'weak' (Pritchard 2005). If the weak survive they may reproduce, thereby passing on their weakness. The aim of eugenics is to actively manage human reproduction to encourage the fittest to reproduce and discourage the least fit from reproducing (Pritchard 2005: 82). Women with disabilities are more likely than non-disabled women to be considered as being at risk of producing children with disabilities.

> The heredity nature of some (but not most) disabilities may call into question the right of disabled parents to produce a baby that may carry their genetic code, disability included ... Certain drug treatments can also increase the risk of fetal abnormality. (Prilleltensky 2003: 23)

Megan Pritchard (2005: 81) poses the question 'can there be such a thing as a "wrongful" birth?' She is interested in the development of pre-natal testing for impairment and the increase in courts 'around the world' hearing cases of 'wrongful birth'. To explore this Pritchard examines the eugenics movement. Drawing on the work of Pilnick (2002), Pritchard argues that although eugenics had little impact on policy in the United Kingdom, in the United States it led to the development of the Immigration Restriction Act and involuntary sterilization laws. Hubbard (1997) estimates that by 1935, 20,000 people had been involuntarily sterilized in an attempt to prevent what the state saw as 'wrongful births' (Pritchard 2005: 82).

There were also institutional arrangements in New Zealand from the late 1800s that segregated and controlled the 'physically anomalous body' (Sullivan 1995). Martin Sullivan (1995: 12, italics in original) explains:

> In 1903 W.A. Chapple, a New Zealand born and bred politician and surgeon, published *The Fertility of the Unfit*. In it, Chapple advocated negative eugenics as a panacea for the colony's ills. ... The solution lay in ... encouraging the *fit* to have more children and by sterilizing the *unfit*.

The 'unfit' included those with 'physical defects' such as the 'deaf, dumb, blind, epileptics, paralytics, cripples, debilitated, and deformed' (Chapple 1903: xii cited in Sullivan 1995: 12). Chapple recommended that such people undergo compulsory sterilization. Physically disabled women of child bearing age, and the wives of disabled men, according to Chapple, ought to have tuba ligature. Chapple's book was widely read and supported (Sullivan 1995).

The Eugenics Education Society of New Zealand was founded in 1910. Amongst its members were some influential and well-resourced people including politicians, academics, and F. Truby King who was Medical Superintendent of Seacliff Mental Hospital. King 'frequently voiced concern at the country's declining birth rate and the physical and moral degeneration of Empire and country' (Sullivan 1995: 13). In August 1913 King gave an address at the National Congress in London entitled 'The New Zealand Scheme for Promoting the Health of Women and Children'. King (1913: 3) stated that the first aim of the Society was:

> To uphold the sacredness of the body and the duty of health; to inculcate a lofty view of the responsibilities of maternity and the duty of every mother to fit herself for the perfect fulfillment of the natural calls of motherhood, both before and after childbirth, and especially to advocate and promote the breastfeeding of infants.[4]

In 1915 His Majesty the King conferred on the Society the honour of being titled 'The Royal New Zealand Society for the Health of Women and Children'. From around 1925 the Royal New Zealand Society for the Health of Women and Children (Inc.) became more commonly known as the Plunket Society.[5] The Plunket Society aimed at producing 'healthy' children and making infant mortality in New Zealand one of the lowest in the world. Deem and Fitzgibbon (1953: foreword) in *Modern Mothercraft* the official handbook of the Plunket Society claims: 'The work of the Plunket Society has had a profound effect in laying the foundations for a healthy nation, and there is growing evidence, too, that other countries are finding the system of great help and value.' Sullivan (1995: 13) argues convincingly that although the stated goal of the Plunket Society was to reduce maternal and infant mortality, King believed that disciplined and highly regulated bodies created moral and normalized citizens. The objective was, therefore, to

> Reduce moral degeneracy by producing fit, healthy, whole, complete, working Truby King babies by training the nation's mothers in techniques of mothercraft. Mothercraft consisted of prescriptive norms of mothering, body technique, corporeality and character structure which constitute and coalesce in a new national icon — the *Truby King baby.* (Sullivan 1995: 13–14, italics in original)

Over the years it has been widely stated that millions of New Zealanders 'owe a debt of gratitude to Sir Truby King' (Snowden and Deem 1951: 8). Yet King's regime did much to put women under surveillance and constructed a cultural hegemony around 'perfecting' mothers and babies that had not existed in the same way prior to his interventions.

Some mothers in New Zealand in the 1950s experienced The Plunket Society's regimes as rigid and harsh (but see Bryder 2003 on New Zealand mothers displaying more agency than might be expected in relation to the Plunket Society). There was little allowance made for women's social, cultural, or embodied differences. For example, all women were told to breastfeed no matter what. I recall my mother saying that the Plunket nurse told her she had to breastfeed my sister (for various reasons my mother wasn't keen on this) unless she had a letter from her doctor. My mother refused and so eventually the nurse bottle fed her. The nurse, though, was angry with my mother and so wouldn't bring the baby to her for feeding. My mother also told me a story about my aunt. She had not wanted to breastfeed because her nipples were extremely sore, but the nurse made her. The baby was forced to drink milk and blood.

Not only was breastfeeding (or natural feeding as it was often called) heralded as ideal for *all* mothers and babies but also there were very strict instructions as to how and when feeding should take place (see Deem and Fitzgibbon 1953: 52–71). The suggested 'Routine Day for the Nursing Mother and her Baby' as outlined below in the Official Handbook of the Plunket Society (Deem and Fitzgibbon 1953: 63) is testimony to this regimentation — the social inscriptions — that mothers (and babies) in New Zealand faced during this era.[6] The aim was to create physically healthy and fit [read: not disabled] mothers and babies.

'A discourse involving the norms of corporeality, technique, measurement and judgement quickly grew up around King's notion of the physically fit infant' (Sullivan 1995: 14–16; also see Ettorre 2000 on 'reproductive genetics, gender and the body'). A wide-ranging infrastructure of surveillance emerged which subjected the mothers and babies to constant scrutiny. One of the forms that this scrutiny and policing took was the issuing of a *Plunket Book* to all mothers from 1924 onwards (Sullivan 1995: 14). These books are still issued to all mothers and infants today. Plunket Books contain a weight chart, room for mother to record baby's progress or lack of progress, and space for the Plunket nurse to record her observations and suggestions about feeding and care. It can be nerve-racking for mothers waiting for the Plunket nurse to visit; waiting for her to measure and weigh the baby, wondering what she will enter into the book — an entry that will be recorded for posterity.[7] When I was visited by a Plunket nurse in June 1995 she wrote in the Plunket Book: 'You have a healthy boy fully breastfed on demand. Mum — have three good meals a day; lots to drink; rest when you can!' Along with this written advice she told me that I *should* put a toy in the basinet for the baby to look at. The Plunket Book and visits by Plunket nurses can be sources of support for women who are caring for babies but they can also function to survey and normalise the corporeal and emotional behaviours of mothers and babies.

Currently, not only is a record kept of the baby's progress after birth, but also a record is kept of pregnant women's health, well being and 'growth'. It is common during antenatal visits for pregnant women to have their urine analysed for the presence of sugar and albumen (protein) and their blood pressure measured. Then, the fundus (top of the uterus) is measured, the fetal heart is checked and the abdomen is palpitated in an attempt to determine the baby's position. Women used to be weighed but this is less common today. This information is noted by the General Practitioner or Midwife in their records. It is also entered on a card, referred to as a 'co-operation card' in the United States, which pregnant women themselves keep (Kitzinger 1989: 48–49). Women bring this card with them each time they have an antenatal check-up. This record functions in a very similar way to the Plunket Book. The record can be seen as a surveillance tool which works to regulate and normalise the bodies of pregnant women. For physically disabled women whose bodies do not fit the norm this kind of regimentation and record-keeping could be a continual reminder of their embodied 'difference'.

Māori women were also continually reminded of their embodied 'difference'. Linda Bryer (2003) in her history of the Plunket Society 1907–2000 argues that the relationship between Plunket and Māori is complex. The Plunket Society, for many years, had a monocultural image. The New Zealand Health Department and Plunket came to an agreement early on that Department rather than Plunket nurses would address and manage Māori infant health. Māori women were permitted to use Plunket's services but few did so. Groups such as the Women's Health League, although not as well resourced as Plunket, were more popular with Māori. After the Second World War, however, Māori urbanization gained momentum and the segregated services could not be maintained. Plunket has had to work at combating its racist image. These days not only does it have to prove that it offers a credible service to Māori but also for a range of other ethnic groups residing in an increasingly multicultural New Zealand (Bryer 2003).

To return to the issue of disability, over the years the influence of eugenics and the discursive space it occupies has waned (Pritchard 2005: 82) although certainly not disappeared (see Pilnick 2002 on how developments in genetics in recent years have revived this discourse). Many still do not expect disabled women to give birth and to become mothers. One of the major themes that emerged in Prilleltensky's (2003) interviews was that the health-care system in Canada was unable to deal effectively with the dual factors of pregnancy and disability. It is common for women without disabilities to feel a loss of control of their bodies when dealing with health and medical professionals. Women with disabilities are likely to have more frequent contacts with the health-care system over the course of their pregnancy, a system which has long tended to label them as abnormal. Prilleltensky (2003: 29) notes:

> while the medical system has a long history of managing illness and disability, it is literally in its infancy when it comes to dealing with the reproductive issues of women with disabilities. Consequently, women with disabilities are often 'groping in the dark' as they attempt to gain information about pregnancy in the context of disability.

Many physically disabled women though are no longer prepared to accept the assumption that their disability is a defect. There is greater acceptance that disability is a social construction rather than a medical condition and in many instances there is no reason why a woman with a physical disability ought not have a baby. Some physically disabled women are now beginning to share their stories about pregnancy and motherhood in an attempt to encourage other disabled women to have children (Rogers 2006). A strong focus in these stories is the physical and mobility challenges presented by various material environments for disabled pregnant women and mothers.

MATERIAL SPACES AND MOBILITY

While able-bodied women sometimes find the final few months of their pregnancy challenging in relation to feeling heavier, clumsier, less mobile, and more likely to topple off-balance this tends to be accentuated for pregnant women who have a physical disability. Often they face mobility challenges earlier in their pregnancy (see Butler and Bowlby 1997 on disabled people's experiences of public space). These mobility challenges occur not just in public spaces but also at home (see Dyck 1995 on disabled women's 'hidden geographies').

One participant in Rogers' (2006: 205) study, Sally, who has a spinal cord injury, found that she was unable to stay as upright in her chair. She kept sliding down. Others experienced more difficulty transferring or being lifted from one space to another. For example, Nadine, who has limb-girdle dystrophy, had problems with her balance because of her larger pregnant stomach. She found it difficult transferring from being seated to standing, and getting in and out of bed and the bathtub. In the third trimester she stopped driving. Natasha, who also has limb-girdle dystrophy, comments that pregnancy exacerbated the difficulty that she has getting up from chairs and couches. In short, a number of the 90 respondents in Rogers' study noted increased mobility difficulties during pregnancy. It is vital, therefore, that these women receive the assistance and the necessary equipment they require whether it be a trapeze lift, step stool, bathtub seat, power wheelchair, walker, or 'reacher' to make adaptation to pregnancy as easy as possible.

Sometimes even maternity wards are not equipped to cope with disabled women's needs (see Chouinard 1997 on ableism). Grue and Tafjord Lærum

(2002), referring to their study of 30 physically disabled mothers in Norway, explain 'several women told us that the maternity ward had no toilet accessible for wheelchair users or other facilities that made it possible for them to nurse their newborn child. This fact may reflect the view that disabled women are not *expected* to be mothers' (Grue and Tafjord Lærum 2002: 673). Within medical discourse disabled women are still often perceived as patients rather than as women capable of giving birth.

Patricia McKeever *et al.* (2003) investigate the experiences of 11 mothers in a large Canadian city who use wheelchairs and scooters for full-time mobility. They conclude 'mothering with a disability consists of embodied and emplaced practices. Women often felt "out of place" as mothers given the relationship between mothering discourse and mobility devices' (McKeever *et al.* 2003: 179). Women with disabilities are often regarded as incapable of looking after children (Gill 1997; Moss and Dyck 1996). However, not only do these women face disapproval from others, especially if it is thought that their children may have to look after them (Prilleltensky 2004; Stables and Smith 1999) but also it can be problematic finding the equipment required to aid disabled women's mobility. Kent (2002), for example, explains that because she is blind, pushing a baby stroller in front of her was not an option. It was unrealistic to rely on her guide dog to warn her about curbs or obstacles at such a distance. Kent called a number of manufacturers in an attempt to find one that produced a stroller with a reversible handle so that she could pull the stroller behind her instead of pushing in front of her into 'the great unknown'. She eventually found one.

Some community groups have recognized the specific needs of disabled mothers and fathers and set up support systems. For example, 'Disability, Pregnancy and Parenthood *international* (DPP*i*) is a small UK-based registered charity, controlled by disabled parents, which promotes better awareness and support for disabled people who are considering pregnancy, during pregnancy, and as parents (see Disability, Pregnancy and Parenthood international, 2005, available <http://www.dppi.org.uk/>, accessed 27 July 2005). But much more remains to be done. Structural arrangements (by way of government prioritizing disability, policies on disability, social services, support agencies, funded assistance) need to be in place to facilitate mothering for those with physical disabilities. Environments need to be made accessible to all (Gleeson 1999) but this extends beyond just the materiality of environments to also include discursive and emotional environments.

CONCLUSION

Pregnancy and parenting is a time of joy, and sometimes sorrow, for many women but for women with physical disabilities the stakes can be higher. Disabled women can face different sets of issues and may need extra support. In relation to the emotional spaces occupied by physically disabled pregnant women they are likely to experience both societal approval and disapproval in different spaces at different times, from different people. They are also likely to experience a vast range of emotions as they grapple with the reactions of loved ones, family, friends, acquaintances, and strangers, and with the health-care system, over the course of their pregnancies.

In relation to the discursive space occupied by physically disabled women they are likely to suffer the weight of the heavy historical legacy of eugenics. This movement created an oppressive discursive space for disabled women and their offspring that constructed them as lacking. Eugenics has not entirely disappeared today. There is still a societal pursuit for 'perfect' babies that suggests disabled people are some kind of unfortunate mistake. But not only are disabled women made to question their very own existence, also they are more likely than their able-bodied counterparts to be perceived as being at risk of producing disabled children. 'The hereditary nature of some (but not most) disabilities may call into question the right of disabled parents to produce a baby that may carry their genetic code, disability included' (Prilleltensky 2003: 23). However, disabled women have resisted this societal quest for bodily 'perfection' and instead taken up their rights to become mothers. Like any individual physically disabled women have rights to lead autonomous lives. They also have the right to make important 'life decisions' as they see fit.

Finally, the chapter looked at some of the disabling effects of material environments including homes, clinics, and birthing wards. Some disabled pregnant women and mothers require assistance to carry out many of the physical tasks associated with parenting because all too often spaces and places are constructed in ways that assume people are able-bodied. Hopefully the chapter has provided yet another window on maternal bodies, gender and space. The subjectivities 'disabled woman' and 'mother' don't always sit comfortably together. Emotional, discursive and material spaces and places constitute and in turn are constituted by the bodies that inhabit them. Women's choices and actions are both enabled and constrained by these spaces and places they inhabit. 'New' selves as mothers simultaneously emerge and are quashed as physically disabled women assert their right to get pregnant, bear and rear children.

Too often disabled mothers are (re)presented as 'bad' mothers. The next chapter also discusses (re)presentations of a 'bad' mother, a pregnant woman known in the media as 'Nikki' who wanted the birth of her baby to be filmed for inclusion in a pornographic movie.

5 A Pornography of Birth

Crossing Moral Boundaries

Birth, in 'the West',[1] has long been represented as a critical time in a woman's life when she needs to be introspective, focused completely on the labour process and on ensuring her baby's safe passage into the world. Birth is also widely considered to be a private affair. Although health and medical professionals are sometimes present to assist, the only other people who usually share the birth process with mothers are their nearest and dearest. Occasionally friends (usually women) are involved but birth is not considered to be a 'spectator sport'. Birthing women have long been expected to 'perform' (Butler 1990) birth in particular ways in particular contexts. The regulatory gender, social and cultural practices that surround birth are established through repeated performances of expected behaviors. Birth is widely understood to be an intimate and profound event. Shelia Kitzinger (1989: 7), a well-known pregnancy and childbirth educator, writes of birth 'It is exciting, awe-inspiring and deeply satisfying.'

Chapter 5 relays a story about a pregnant woman in Hamilton, New Zealand,[2] who is known by the media only as Nikki (her screen-name). Nikki wanted her birth to be 'exciting, awe-inspiring and deeply satisfying' but one expects not quite in the manner imagined by Kitzinger. Nikki wanted be filmed giving birth as a prompt for strangers' sexual excitement and gratification — she wanted the birth to be included in a pornographic movie. The story of Nikki aired in July 2002 when a New Zealand *60 Minutes* television documentary featured her explaining that she planned to be filmed giving birth for a movie entitled *Ripe*. The film is about a woman's — Nikki's — sex life during pregnancy and the final scene, her giving birth, was to be the 'climatic moment'. Immediately following the television documentary a media furore erupted.

The aim of this chapter is to draw on the story of Nikki and pornographic film maker Steve Crow's[3] quest to have a birth filmed for a pornographic movie to illustrate that certain sexual acts rouse anxieties and even disgust (see Church Gibson 2004 and Rodgerson and Wilson 1991 on pornography, gender, and feminism). The moral boundary between what is considered 'normal' and what is considered 'perverse' is constantly struggled over and is temporally and spatially specific. This pornography of birth shows that what counts as moral is tied up with issues of gender, sexuality, class, race and so on, but also with 'geographical objects of space, place, landscape, territory, boundary and movement' (Cresswell 2005: 128). Tim Cresswell (2005: 128) explains 'A moral geography, simply put, is the idea that certain people, things and practices belong in certain spaces, places and landscapes and not in others.' This chapter shows how Nikki, through media discourse, was constructed as a person who belonged in certain places and spaces (brothels, strip clubs) but not in others (hospital birthing wards). The media represented Nikki as immoral but this morality turns out to be based on a very contingent set of societal rules and expectations.

The chapter is divided into five sections. The first section reviews briefly recent research on moral geographies. My reading of this research on moral geographies is filtered through a specific interest in Julia Kristeva's notion of abjection and a more general interest in feminist poststructuralist theorizing on bodies. Second, the methodological process undertaken to conduct the research, namely collecting and critically reading media reports, is discussed. In the third section readers are introduced to Nikki and filmmaker Steve Crow, then provided with information about the unfolding of events surrounding the filming of the birth of Nikki's baby's for inclusion in a pornographic film. Fourth, a critical reading of media reports on Nikki and Crow's plans to film is presented in order to argue that the moral boundary between what is considered 'normal' and what is considered 'perverse' is constantly struggled over. Unwritten rules and regulations govern what is deemed (in)appropriate behavior for particular bodies in particular spaces producing 'a changing sexual landscape' (Weeks 1995; 2003) and constantly shifting 'moral geographies' (Smith 2000). The chapter concludes that Nikki parodies the purity and naturalness of pregnancy, birth, and child rearing by constructing the space of the birthing ward as explicitly sexual. It is important to question what counts as moral and immoral and to show how morality is infused with geographical notions of space, place, access and boundaries.

'MORAL GEOGRAPHIES'

Philip Hubbard (2000: 198), drawing on the work of David Smith (1997), explains the term 'moral geographies' has emerged in the past decade to describe empirical research into those aspects of sociospatial ordering which 'invite a moral reading' (Smith, 1997: 587; see Smith 1997 for a useful review of the emergence of 'geography and ethics' and 'moral geographies'; also see Matless 1994; Setten 2004; Smith 2000). Smith (1997: 587) argues an important role for geographers is to 'take up where most philosophers leave off: to examine the contextual thickening of moral concepts in the particular (local) circumstances of differentiated human being[s].'

A good example of this kind of research, and one that is particularly relevant to this chapter, is Hubbard's (1998) examination of a red-light district in Birmingham, UK (also see Hart 1995). Focusing on community protests against prostitution in Birmingham, Hubbard highlights 'the way that moral narratives and discourses were deployed by protestors in their attempt to construct an idea of community predicated on the exclusion of "immoral" sex workers' (Hubbard 1998: 55; also see Hubbard 1997). Hubbard's work usefully illustrates that understanding sexualized acts and spaces is multifaceted and contradictory since (hetero)sexuality does not stand alone but is entangled with gender, race, ethnicity, social class, age and so on. Nikki is not a prostitute but she is involved in the sex industry and does therefore, inhabit particular 'moral landscapes' (Setten 2004) in which a range of moral discourses about 'appropriate' gender, sexual, and class behaviors apply.

Cresswell (2005: 128) explains that Felix Driver (1988) was the first to use the term 'moral geographies' but its lineage can be traced through the work of people such as Chris Philo (1987), David Sibley (1981; 1995) and indeed Cresswell himself (1996; 1997). These works pay attention to the various ways in which some people become constructed as outsiders and as 'out of place'. Cresswell (2005: 128) notes 'They are out of place because they do not fit into an already established (even if only temporarily) set of expectations about the link between geographical ordering and behaviour.' People such as the disabled (Kitchin 1998b), children (Philo 1992; Valentine 1997), young people (Skelton and Valentine 1998), gays, lesbians and bisexuals (Bell and Valentine 1995; Brown 2000), the homeless (Daly 1996), gypsies (Sibley 1992), prostitutes (Hubbard 1998) and many Others are constituted as having no place within 'normal' society.

Cresswell (2005) explains that the aforementioned geographical works on moral boundaries, outsiders, and Others have often been variously informed by Erving Goffman's contributions on stigma (Goffman 1968), Michel Foucault's research on ordering and discipline (Foucault 1977; 1980), Pierre Bourdieu's investigations on 'doxa' and 'habitus' (Bourdieu 1990), the Chicago School sociologists' earlier and later works on 'moral orders' and 'deviance' (Park and Burgess 1925) and psychoanalysts' work on 'object relations' theory (Klein 1990).

My reading of, and drawing on, contributions on 'moral geographies' has been informed mainly by object relations theory which explores 'relationships between the self and the social and material world' (Sibley 1995: 7). I've been especially influenced by Julia Kristeva's work on abjection (Kristeva 1982), a particular branch of object relations theory, which has helped further my understanding of some of the anxieties surrounding the boundaries between the pure self and defiled other (see Sibley's 1995: 5–11 account of object relations theory and abjection).

I have explained the notion of abjection elsewhere (see Longhurst 2001b: 28–32; also see Sibley 1995: 8–9) but in short, abjection can be defined as meaning 'to expel, to cast out or away' (McClintock 1995: 72). Kristeva uses the notion of abjection in her book *Powers of Horror* (1982) to illustrate the significance of various personalized bodily horrors for subjects (as they exist within particular cultures) in relation to the boundaries and orifices of the body. Kristeva examines the conditions under which the clean, decent, proper body is demarcated. The cost of this body emerging is what she calls abjection.

> Abjection is the affect or feeling of anxiety, loathing and disgust that the subject has in encountering certain matter, images and fantasies — the horrible — to which it can respond only with aversion, nausea and distraction. Kristeva argues that the abject provokes fear and disgust because it exposes the border between self and other. (Longhurst 2001b: 28; see also Grosz 1994: 192)

Many of these insights on abjection are anchored by Mary Douglas' work on boundary rituals and dirt. Her now famous dictum that dirt is essentially disorder, it is 'matter out of place' (Douglas 1966: 25) illustrates that nothing in itself is dirty, rather dirt is that which does not belong. 'The idea of dirt takes us straight into the field of symbolism and promises a link-up with more obviously symbolic systems of purity' (Douglas 1966: 35). Nikki, as is illustrated in the remainder of this chapter, is constructed through discourse as dirty, impure, and dangerous. Her actions prompted, for some, feelings of fear and disgust and abjection. One newspaper article is entitled 'Repulsive sleaze' (*The Press* 21 October 2002). The idea of filming Nikki's birthing body engendered in doctors, midwives, health board officials, Members of Parliament including the Minister of Health, journalists, and the general public feelings of anxiety, loathing and disgust. Consequently, there were numerous attempts by these people to reinstate Nikki's maternal body as clean and proper (this is discussed in more detail later in the chapter).

Moral geographies, and 'heretical or immoral geographies' (Cresswell 2005: 129), that is, geographies concerned with social exclusion, 'outsiders', Others, transgression and people who are said to be 'out of place', can be read alongside social and cultural feminist work on embodiment, and in particular on abjection. This work has informed the research. Notions of disgust, disorder, impurity and dirt are useful for thinking through Nikki's 'social geography of birth'.

A NOTE ON METHODS

I have been engaged in research on pregnancy for approximately 15 years. Over that period I have collected a huge array of scholarly articles and 'popular' materials not only on pregnancy but also on birth. These include stories in lifestyle, health, and women's magazines, but also representation of birth on television, in documentaries and film. I have not drawn specifically on these data in this chapter but they did provide a broader context for the research.

The main data presented in this chapter is from two *60 Minutes* New Zealand television documentaries and newspaper articles on Nikki. The first television documentary screened in July 2002, the second (after the birth of Nikki's daughter) on 14 November 2002. Copies of these programmes were requested and received from TV3 so that I could watch them a number of times. During this period July–November 2002 I located 20 newspaper articles which appeared in *The Press*, *The Dominion Post*, *New Zealand Herald* and especially the *Waikato Times* (the local newspaper in Hamilton, where Nikki was residing at the time). Many of the articles were tracked through the website 'nzoom.com news'. In addition to watching the video and reading newspaper articles, I listened to several talk-back radio stations including Newztalk ZB to gauge the public's reaction to the story as it unfolded. While I do not draw specifically in the chapter on these data from talk-back radio, they were useful for gauging reactions to the story from at least some members of the public.

I considered supplementing the data outlined above with data from interviews. I wondered if it might be useful to talk to Nikki and/or Steve Crow, director of Vixen Direct. However, both Nikki and Crow were claiming in media reports that they had been plagued by reporters. 'Crow says he's tired of all the media attention' (Holt 19 October 2002: 19). In the same newspaper article the reporter writes:

> *The Times* attempted to talk to Nikki at Waikato Hospital. She covered her face with her hands when she saw a photographer and reporter in the reception of ward 55 requesting a meeting. 'I'm not allowed to give any interviews,' she said as she was led away by a nurse. (Holt 19 October 2002: 19)

Under the circumstances it seemed unlikely that I would be successful in gaining an interview with either Nikki or Crow, nor did I want to add to the current difficulties they were facing so I decided against attempting to interview them. Also, in the final instance, I am most interested in the way in which Nikki and Crow are discursively constructed through the media and the power relations inherent within these constructions. I was not concerned with attempting to secure *the* truth of the matter from the *real* people involved in the saga.

In analyzing the data I focused on issues on power, regimes of truth, and on how moral narratives and discourses are deployed by various individuals and institutions to (re)present certain images of women, birth and pornography. Stuart Aitken's (2005) work on textual analysis, especially on poststructural readings of texts (248–49), is informative. Aitken argues that poststructural text analysis involves treating texts (in this case, television and newspaper reports) as social products and recognizing that each text refers in some way to another text. He concludes:

> At the level of textual representation, a poststructural perspective questions not only what is known, but also how it comes to be known. It is based upon the simple perspective that nothing in the world is fixed or immutable, that things are grounded on moving foundations. And perhaps most importantly, poststructuralism questions the basis of any method that assumes a structure of signification and understanding that is not politically based. (Aitken 2005: 248–49)

I took this approach to analyzing the data looking for the emergence of particular themes in the *re*-presentations. One theme that emerged was moral outrage. I also looked for complexities and complications that contradicted and problematised this theme of moral outrage. I was interested in interrogating the story-telling, but also audience responses, and my own responses to the stories.

BIRTH RITES: INTRODUCING NIKKI

Tristan Taormino (2002) in New York's *The Village Voice* comments on pregnant Hollywood celebrities who she describes as 'Hot Mamas':

> Did you see all the big bellies on TV a few weeks ago? Watching this year's Emmy Awards, I was struck by the number of gorgeous pregnant women on the red carpet. From *Malcolm in the Middle*'s Jane Kaczmarek to *Sex and the City*'s Cynthia Nixon, Hollywood mommies-to-be were stunning and sultry. Not oh-you're glowing-like-the-Virgin-Mary goddesses, but really hot-to-trot vixens. (Taormino 2002)

Taormino continues that positive portrayals of pregnant women as sexual is a 'good thing' because for too long, many people have treated pregnant women as 'utterly asexual, as fragile, [and] in "a delicate condition".' While pregnancy in 'the West' has undergone something of a make-over in the past decade (Longhurst 2005a and Chapter 3 this volume) it seems that birthing has not. While some pregnant women are adopting fashionable and sometimes revealing styles of dress as a way of resisting constructions of the pregnant subject as modest, respectable, domestic, asexual, and private *re*-presentations of birthing women do not seem to have changed as radically.

Birth, still for the most part, tends to be seen as a special time in a woman's life when she needs to be focused wholly on the new life that she is about to bring into the world. Birth is widely considered to be a personal journey. Although midwives and medical practitioners may be present to assist the mother and baby the only other people who tend to be present or involved are the birthing woman's nearest and dearest. There are societal expectations that birthing will be enacted in particular ways. Regardless of whether it be a 'natural' birth, a pain-assisted birth, a forceps delivery or a caesarean section the expectation is still that birthing women ought to behave in culturally and gendered 'appropriate' ways. Nikki's plan to be filmed giving birth for a pornographic movie was not seen by most as an 'appropriate' way to birth.

Nikki worked 'on and off' as a stripper for several years (Holt 24 October 2002). Her family lives in Northland, located in the upper North Island of New Zealand, although Nikki is reported to be a former Huntly College student (Holt 24 October 2002). Huntly is a small working class town with a population of approximately 7,000 residents. It is located on State Highway 1, 35 kilometers north of Hamilton. The landscape is dominated by a massive coal-fired power station (see Huntly, Welcome to our town, 2005, available <http://www.huntly.net.nz/>, accessed 23 March 2005). In the United States Nikki might disparagingly be referred to as 'white trash' — she is an Other.

Only one newspaper, the *Waikato Times*, in an extended weekend feature, offered any background information on Steve Crow. He is described as holding an honours degree in marine biology from Auckland University. Crow says:

> he fell into the industry 'quite by accident'. He was running a computer company and, while on a business trip to Taiwan in 1993, saw an opening to import porn-based computer games to New Zealand. 'It all went from there'. (Holt 19 October 2002: 19)

The article on the saga includes a large (22cm × 25cm) colour photo of Crow (from the waist up). He is standing, stern-faced, arms folded, in a room walled with pornographic videos. Crow is wearing a brown collarless sweat shirt, an earring in his left ear. He is of stocky build and has a bald/shaven head (Holt 19 October 2002: 19). The caption under the photograph reads 'STEVE CROW: "There was two hours of talk-back this morning. They want me shot. Everyone's missing the point. It's a civil liberty issue".'

A smaller (7cm × 6cm), head and shoulder photo of Nikki is included in the same article. She has long brown hair, thin penciled eyebrows, dark eye-liner, dark lip-liner, and is smiling. Her shoulders are bare except for a thin pink strap of a halter-neck top. The caption under the photograph reads 'CENTRE of attention, Nikki.'

Nikki's sexual/employment practices (making pornographic movies) cannot be regarded simply as an explanatory variable for the moral outrage that ensured during the salacious media frenzy about her birth plans. Her social class, the way she speaks, her comportment, her 'look', her age

(22 years), her and her family's place of residence, are also factors in the way her corporeality is read by the media and the public. Notions of maternal morality can be seen to intersect with factors such as gender, age, and social class.

After the documentary about Nikki screened, the media promptly began to debate appropriate behaviors for pregnant women and the rights of unborn children. Nikki was dubbed the 'porn mum' by the media. Questions were raised as to whether it was lawful to film the birth for a pornographic film. The case was taken by Child, Youth and Family (CYF), a government department, to the High Court. CYF sought to ascertain whether they should be given guardianship of the unborn child and whether Nikki could be filmed giving birth for *Ripe*. Justice Heath declined CYF's application for wardship and instead appointed Nikki an agent of the court. The filming could proceed but it was ruled that the filmmaker could not use any shots of the unborn baby, or images of the newly born baby, in the movie. This would breach the UN Convention of the Rights of the Child (which has been signed by New Zealand). Nor were any shots of hospital staff, patients, or visitors to be included.

At this point, many individuals and institutions urged Nikki, for the sake of her child, to abandon her plan for filming but she was adamant she wanted to proceed. Government Health Minister Annette King entered the debate and used her statutory powers to ban the filming on District Health Board Premises. King directed the Waikato Board to say no to the filming (see King intervenes over porn birth, 2002, available <http://onenews.nzoom.com/onenews>, accessed 16 February 2005; McLeod 2002). The Board had initially refused to allow any filming, but changed its mind after the court ruling that filming was lawful. Crow and Nikki began to talk to private health providers in an effort to circumvent a ban on him using public hospitals for filming pornography. Nikki was angry about the ban on cameras in the delivery suite saying this was unfair because many other women take videos and photos of their birth for sentimental reasons and to show family and friends. Nikki too claimed that she wanted to show photos of the birth of the first grandchild to family who lived in Northland.

During this period of legal battles, 35 week pregnant Nikki was admitted to hospital with complications that possibly required the birth to take place in a hospital with facilities for an emergency caesarean section. Crow announced on 30 October 2002 that 'he was pulling out of the deal to film the birth' (*Waikato Times*, 31 October 2002: 1) on account of Nikki's health problems (it looked likely that she would deliver by caesarian section). Crow's decision to pull out may also have been on account of the ongoing public outcry about the proposed movie. On 30 October various newspapers across the country announced 'It's a girl! Porn actress now a mum' (Holt 30 October 2002, 1). Following the birth of Nikki's daughter who was named Brianna, the media frenzy died down.

MORAL OUTRAGE: SULLIED SPACES

At stake in the making and screening of pornographic films about pregnancy and birth are questions about the conceptualization of pregnant, birthing, and babies' bodies. When birthing the borders of a woman's body change rapidly. One self 'splits' itself into two or more. Another human being crosses a border — a bodily border — to enter a new space that is external to the mother's body. Whether the body is born vaginally or by Caesarean section the birth nevertheless marks the point at which both mother and infant start to become individuated subjects. At first they are still attached by the umbilical cord. Breastfeeding which involves the infant literally ingesting the body of the mother may continue for some time. Inceasingly, however, mothers and babies separate and birth marks the beginning of this separation. Birth is also the first time that the mother discovers what her baby looks like (hair, face shape, skin, eyes), smells like, and feels like pressed to the outside of her body.

'Coupling' pregnancy and especially birth with sexual gratification challenges mainstream notions of pregnant and birthing women as modest, 'motherly', and focused completely on their infant. Becoming mothers must not 'flaunt' their sexuality even though (or maybe, because) the pregnant, and especially the birthing body is a body that is [assumed to be] clearly marked as having participated in sexual intercourse (Longhurst 2000). Nikki's transgression, therefore, prompted something of a moral panic (see Holt 12 October 2002 and 14 October 2002).

An editorial in the Christchurch newspaper *The Press* (21 October 2002) claims: 'Views on censorship and pornography vary widely in our society, but this project should be universally condemned as beyond the pale.' Adjectives used to describe the 'porn birth' in 20 articles that were analyzed include 'outrage', 'anger', 'an affront to other mothers', 'an insult to health professionals', 'degradation', 'bizarre saga', 'repulsive sleaze', 'sordid commercialism', 'distasteful', 'abhorrent' and 'repugnant' (see newspaper articles and Internet sources listed in the bibliography under Holt and *Waikato Times*) (see Figure 5.1).

There are very few public expressions of support for Nikki's actions in the media. Rather ironically, in one newspaper article (Holt 19 October 2002: 19) even filmmaker Crow is reported as saying, 'Personally, I don't agree with it' referring to Nikki's decision to star in the film. However, Nikki comments that Crow has been 'absolutely wonderful' and that she 'couldn't ask for better support' (Holt 26 October 2002: 3). In the same article Nikki also praises the Waikato Hospital staff who she said had really looked after her. One journalist, while not supporting Nikki's decision, did at least allow for the possibility that Nikki might be a 'good' person: 'She [Nikki] aspires to be a good mum and despite the unpromising start, who's to say she won't get there?' (*New Zealand Herald* 16 October 2002). Clearly there are people who support the making of this film (after

THE DOMINION POST, 14 OCT 2002, Edition 2, Page 4.

Doctors angry about role in porn birth

By: BETTS Marianne

DOCTORS are angry that decisions over a birth being filmed for use in a pornographic movie might fall on the shoulders of a medical consultant.

The Waikato district health board has decided to all... consultant had to intervene in the bi...

Page 1 of 1

WAIKATO TIMES, 30 OCT 2002, Edition 2, Page 1.

It's a girl! Porn actress now a mum

By: HOLT Monica

...Nikki gave birth at Waikato Hospital this morning -- a month prematurely and ...ture the moment.

...Waikato Times today to say the girl weighing 2862g (6lb5oz) ...baby were both well.

STUFF — THE PRESS — OPINION — 21 OCT 2002

Repulsive sleaze

A bizarre saga has been played out at Waikato Hospital, writes The Press in an editori...

It involves a yet unborn baby, the mother -- a former strip... stardom, various members of the medical and ... issue is whether this mother-to-be ... allowed to have the ...

Figure 5.1 Newspapers report outrage at 'Nikki, porn mum'. *Source*: Articles downloaded from Internet sites.

all, according to Crow hundreds of copies of the film were pre-sold), but these people did not voice their concern publicly.

For the most part, as reported in the media, the actions of Nikki and Crow crossed a moral boundary and feelings of abjection ran high (see Kristeva 1982 and Sibley 1995: 8–9 on abjection). 'Good' sex (read: 'normal, natural, healthy, holy' [Rubin 1989 cited in Hubbard 2000: 197] sex for procreation) is seen to be turned into 'bad' sex (read: deviant and perverted sex and for financial gain). What constitutes good and bad, although highly complex, tends to be constructed as 'common sense'. Cresswell (2005: 128) notes 'The word "moral" indicates a fairly contingent set of rules and expectations dressed up as though it were common sense.' After Health Minister Annette King stepped in and banned the filming of Nikki giving birth, King commented that she had felt it imperative that she exercised her 'moral judgment' (see Stuff — National News 15 October 2002, available http://www.stuff.co.nz, accessed 16 February 2005). Eleven days later when Crow pulled out of the deal, King announced 'I am pleased common sense has prevailed. I wish Nikki and baby all the best' (Holt 26 October 2002: 3).

In examining moral judgments as to whether birthing women ought to be engaged in invoking sexual feelings for commercial gain it is imperative to consider the relationship between bodies and space, in this case, a delivery suite in a public hospital. Seeking a court order to stop the filming of the birth of Nikki's baby could be read as an attempt to reinstate the purity of the delivery suite — a space where mother and child meet, bond, and establish a positive and loving relationship. When it was proposed that the delivery suite would become the site of a pornographic movie, lines between purity and perversity (see Weeks 2003: 69–90 on perversity) became blurred. While viewing and shooting pornography might be 'tolerated' at sites that are seen to be deviant such as sex shops, clubs, strip joints, warehouses, porn studios, private homes, it was not 'tolerated' in a hospital birthing ward. One journalist wrote 'it defies belief that anyone could imagine that the real-life birth of a child in a hospital delivery theatre should be thrown open to skin-flick makers — and their subsequent audiences' (*The Press* 21 October 2002). The journalist continued, 'operating theatres' should not be allowed to become 'free porn studios' (*The Press* 21 October 2002). Filming a pornographic film in a delivery suite hugely disrupted the prevailing moral order. A delivery suite may be an intensely sensuous space, imbued with feelings of heightened pain, emotion, joy and sorrow, but sexual feelings are usually repressed or ignored in this setting. Another journalist wrote 'The idea of a porn movie crew hanging around the local maternity ward, theoretically poised to film someone else's cervical dilation, would distress me utterly' (*Waikato Times* 23 October 2002). Later in the same article the journalist continues:

> He [Crow] and his crew have no place in a hospital, no place around women suffering in labour, no place near newborn babies. As far as I'm concerned it would be an affront to mothers forced to use the same hospital that their families should be near him, and an insult to health professionals to have to tolerate his presence. (*Waikato Times,* 23 October 2002)

One possible reading of this quote is that it expresses a fear of contagion. Both Nikki and the film crew would be 'out of place' in the maternity ward. Furthermore their very presence presents a risk of sullying other mothers and their families in the vicinity. The hospital delivery suite is widely considered to be a space in which health, medical, and surgical care are offered. As such, although hospitals are sites of 'real' flesh and blood bodies, they are imbued with discourses of science and rationality. They are spaces that must be kept as free as possible from all kinds of pollutants and impurities (bacterial and moral). Filming a birth to include in a porn movie was seen by many, including journalists, midwives, nurses, doctors, social workers, the Commissioner for Children, the Minister of Health, and talk-back radio callers, to most certainly defile the space. Professor Jane Ritchie, an expert in Psychology and Women's and Gender Studies at the University of Waikato, describes the prospect of filming the birth for a porn movie as 'repugnant'. Ritchie claims 'It is not like she is doing it in New York. This is a small country. That child will face stigma for life' (Holt 19 October 2002: 19). Chief social worker Shannon Pakura also mistrusted Nikki and sought to award guardianship of Nikki's baby to the Child, Youth and Family Services Department.

Like the move towards pregnancy chic, Nikki's and Crow's desire to make a pornographic film about pregnancy and birth was driven at least in part by monetary gain (described by one journalist as 'sordid commercialism' [*The Press* 21 October 2002]). Advertisers and retailers are always on the look-out for new products and new markets. In the 1980s advertisers began marketing beauty, fitness, and health products to young men (Miller *et al.* 1998; Mort 1996). In the 1990s advertisers began seeing pregnant women as offering commercial possibilities, especially white, middle-class, fashion-conscious, pregnant women (Longhurst 2005a and Chapter 3 this volume). In the 2000s Crow identified a gap in the pornographic film industry. Although pornography involving pregnant women is nothing new,[4] Crow claimed that to the best of his knowledge no one in the world had ever before filmed a birth for inclusion in a commercial pornographic movie (*60 Minutes*).[5] Crow comments the porn industry in New Zealand is growing. He offers 16 new titles a month to New Zealand's 480 video libraries. Crow received hundreds of pre-orders for *Ripe*, which will cost NZ$69.95 to purchase (Holt 19 October 2002: 19). Porn actors, he says, make around NZ$30,000–40,000 a month. It is perhaps worth noting, however, that Crow spent NZ$15,000–20,000 on legal fees and so in the final instance maybe stood to make less money from the film, unless publicity boosted sales. Crow vowed to give all profits from *Ripe* to Nikki's unborn child (Holt 19 October 2002: 19).

And yet filming Nikki's labour and the birth of her baby was not just about financial gain. It was about morality, difference and human rights. Crow says 'It's a civil liberty issue ... it's not something I would want my daughter or wife to do, but it is Nikki's right to do it. I think sticking a cigarette in your mouth is disgusting, but I don't tell people not to smoke' (Holt 19 October 2002: 19). Crow mentions his rights and the rights of Nikki but it is also about the rights of Nikki's unborn child, the midwife, hospital staff, members of the District Health Board, CYF, the High Court, the Commissioner for Children, the Health Minister and 'the public'. Many individuals, organisations and institutions felt they had some kind of stake in this issue.

CONCLUSION

The aim of this chapter was to examine actor Nikki's and pornographic filmmaker Steve Crow's desire to have a birth filmed for inclusion in a pornographic movie in order to illustrate that the moral boundary between what is considered 'normal' and what is considered 'perverse' is constantly struggled over and is spatially constituted. This pornography of birth shows that certain sexual acts rouse anxieties, and even abjection and disgust, and that what counts as moral is tied up with issues of class, race, gender, sexuality and so on, but also with space, place, territory, and boundaries. By not complying with 'taken for granted', 'common-sense' understandings of motherhood Nikki parodies the purity and naturalness of pregnancy, birth, and child rearing and constructs the space of the birthing ward as explicitly sexual. In doing this she crosses a moral boundary. Nikki is thought to risk not only her own moral reputation but also the reputation of her baby. What is more Nikki's behavior is seen to pose a risk to the sanctity of the birthing ward and to put the other mothers and their families at risk of being sullied.

Pornography might be 'tolerated' by some and embraced by others, in spaces such as the Internet, private bedrooms or lounges, clubs and other sites of erotic entertainment, but not in hospital birthing wards. Birthing wards are widely considered to be 'moral' spaces. Nikki parodies the purity and naturalness of pregnancy, birth, and child rearing by constructing the space of the birthing ward as explicitly sexual. In doing so she crosses a moral boundary and is thought to risk not only her own moral reputation but also her baby's reputation. To borrow Taormino's (2004) phrase, Nikki is a 'hot mama' but she is a 'mama' who is 'too hot' for most.

Birthing and pornography are both cultural performances — performances that tend to be held apart even though sex is the cornerstone of both. In bringing together birthing and pornography there emerged for some a fear of mixing unlike categories. Nikki transgressed boundaries and various people including a High Court Judge, the Commissioner for Children, the Minister of Health, a professor of psychology, a senior social worker, journalists and members of the public who drew on moral narratives and discourses to discipline her in an attempt to restore respectability and the sanctity of the hospital birthing ward. Nikki was constructed as a threat to 'proper', 'decent', and 'respectable' mothers.

David Smith (2000: 1) notes:

> Human beings have moral values … ideas about what is right or wrong, good or bad, better or worse, in connection with important aspects of life. These values guide our actions, helping us to decide what we should or should not do, how we ought to live; and they provide a basis for evaluation of the conduct of others. They can vary among individuals and groups, and hence from place to place, as part of our world of difference.

Geographies of morality, identity and difference are useful in furthering understanding of people's relationship with places, especially when sex is involved. Sexual performances are constituted within different contexts of moral regulation. The boundary between what is considered 'normal' and what is considered 'disgusting' or 'sick' is constantly struggled over. Particular kinds of sexual practices and sexual identities are deemed appropriate in some spaces but not in others. Different constructions and representations of sexed bodies can provoke feelings of both desire and disgust — abjection — depending on the ways in which these bodies perform in different social and spatial contexts.

Hubbard (2000:197), drawing on the work of Rubin (1984), argues that issues of morality are:

> ...never far away when sex is discussed, serving to naturalize the idea that sex must involve a *meaningful* material and emotional exchange based on procreative sexual intercourse, branding those who indulge in different forms of heterosex as immoral and deviant (or 'perverted') (italics in original).

There already exists in geography, prompted in part by queer theory, a rich literature on sex and sexuality as salient parts of people's subjectivity and on how spaces are powerfully imbued with normative versions of sexuality.[6] Sexual dissidents have challenged normative constructions of heterosexist space and carved out other spaces where it is possible to perform queer sexual identities (e.g. Davis 1995; Johnston 2005; Johnston and Valentine 1995; Valentine 1996).

There is room, however, for much more work that challenges normative heterosexuality, not just work on queer identities, but also on and the relationship between constructions of queer and straight (e.g. Cream 1995; McDowell 1995). There is potential for research on queerness, bisexuality, heterosexuality, pregnancy, birth, romance, passion, desire, disgust, pornography, prostitution and so on in relation to the ways in which (im)moralities are constituted within particular spaces. I look forward to a further opening up of discussion around what counts as moral and immoral and how morality is infused with geographical notions of space, place, access and boundaries.[7]

6 At Home with Birth

In many cultures birthing is an important *rite de passage* and as such it has been the subject of a great deal of anthropological, psychological and sociological attention. The places and spaces of birth, however, have not been the subject of nearly as much attention. It is time geographers addressed this subject — a point reiterated by Sally Abel and Robin Kearns (1991: 825) who argue 'An examination of the role of place in women's experiences of birth is an appropriate task for the geographer.' Scott Sharpe (1999: 91) notes childbirth in the West is an 'intensely political issue'. Questions such as: who is the most appropriate primary care-giver to the birthing woman; and, who ought to have authority to make decisions during the birthing process have been posed repeatedly over not only decades but over centuries. Sharpe (1999: 91) explains that childbirth is also a *geographical* issue: 'Much contention with regard to childbirth in the West has centred around the problem of where childbirth should occur.' Maria Fannin (2003: 515) explains 'Representations of birth are (always) spatial, yet, until recently, popular and academic work on birth tended to elide the significance of the changing spatiality of birth.' Fannin's research on 'Domesticating birth in the hospital' illustrates clearly that birth is both a social and spatial process 'bound up with discursive and material representations that mobilize certain historically and geographically contingent spatial metaphors' (Fannin 2003: 515).

Like Fannin (2003: 513), I want to explore some of the debates around 'domesticity, the body, [and] the politics of reproduction'. More specifically I want to examine home as a site of birth and some of the postcolonial and feminist discourses surrounding home birth. The focus will rest primarily on Aotearoa New Zealand — a site of interesting and complex Māori, postcolonial, feminist home birth politics. Many of the issues discussed though are also relevant to other places.

Abel and Kearns (1991: 825) argue that in New Zealand in 1920 approximately 65 percent of births took place outside of maternity hospitals but 15 years later, only 22 percent of women were delivering their babies at home. In more recent years there has been a slow increase in the number of home births but it is difficult to know the exact numbers. According to McLoughlin (1993: 59) in 1981 there were 253 home births recorded in New Zealand. By 1985 the number of home births had risen to 387. In 1989 there were 900, in 1990 there were 1148, and in 1991 there were 1200 home births recorded by the Home Birth Association. No figures are available on the ethnicity or social class of the mothers who are opting for home births. While the exact numbers themselves may be difficult to determine, what is more certain is that home births represent only a very small proportion of the total number of births in New Zealand. In 2005 it is estimated that only one percent of women had a planned home birth.

Many things affect women's decisions about where to birth. For example, in the early 1990s in New Zealand the political climate surrounding birthing began to change. The closure of small hospitals and the centralizing of obstetric services meant that women had to think carefully about birthing options (Able and Kearns 1991). The closure of a number of small hospitals meant that the distance women in labour must travel to their nearest publicly funded obstetrics hospital had increased. Paul Beere and Lars Brabyn (2006) employ Geographical Information Systems (GIS) to map health accessibility data to determine travel times for birthing women to publicly funded maternity hospitals in New Zealand. It is likely that the increased travel times that have resulted from a drive for economic efficiency may have affected some women's decisions about where to birth (Beere and Brabyn 2006).

Another change that has occurred in recent years is the emergence of private birthing facilities. Many of these facilities are 'designed to de-emphasise the clinical trappings of a delivery room, and appeal to families who might otherwise opt for a home birth' (Abel and Kearns 1991: 826). In other words, some hospitals and birthing centers have attempted to make birth spaces more 'homelike' (see Fannin 2003). This move has in part been underpinned by an attempt to demedicalize birth and allow women to take control of their own birth experience in their own space.

This chapter provides yet another window on maternal bodies — it examines birthing bodies and the importance of spatiality in relation to birth. This is a theme that was in part explored in the previous chapter on 'a pornography of birth'. This chapter views birthing through a different lens. The first section outlines the data used to inform the chapter. The second section examines the colonization and reclaiming of birth for Māori women in Aotearoa. It explores the shifts for various generations of Māori women from home birth, to hospital birth, and back again to home birth. A rich narrative from one participant, Angeline Greensill, highlights some of these issues surrounding the place of birth for Māori women. Third, birth is examined not just as a (post)colonial issue but also as a feminist issue. It is argued that those in support of home birth often represent homes as essentially 'good'. Birthing wards in hospitals on the other hand are represented as essentially 'bad'. However, the notion of home needs to be problematized in such discourses. Finally, following Sharpe (1999) it is argued that a binary division between home and hospital has characterized the politics of childbirth. This binary needs to be deconstructed. Birthing centres represent one way of troubling this binary. The chapter concludes that neither bodies, nor places, are essential. Homes and hospitals do not have essences but are imbued with culture and history. It is important to consider carefully the politics of spatiality before drawing easy conclusions about home birth.

COLLECTING DATA

This chapter draws on an in-depth informal interview with Angeline Greensill who is Māori,[1] of Tainui descent from Whaingaroa/Raglan, Aotearoa New Zealand. I chose to talk with Angeline because between 1971 and 1991 she gave birth to four children in a hospital and three at home. Angeline now has grandchildren and over the years has reflected a great deal on the politics and spaces of birthing in Aotearoa. I invited Angeline to discuss her experiences of birthing, especially at home, including how she felt about being at home, who was present, and what prompted her decision to have her fifth child at home (her first four were born in hospitals). The interview was conducted in my office, audio-taped and lasted just over an hour. The tape was transcribed in full.

The chapter is also informed by informal discussions with a range of people including pregnant women and home birth midwives. On 7 November 2004 I attended an open day at the Hamilton Parents' Place in Hamilton, New Zealand. A series of 'information talks' were offered as part of the open day. I attended a 'talk' entitled 'Birthing normally in a caesarean culture' by home birth Waikato midwife Maggie Banks. Banks is also author of the book *Home Birth Bound: Mending the Broken Weave* (2000). I also read a number of 'newsmagazines' produced by the Waikato Home Birth Association Inc, and critically examined Internet sites that discuss various options for birthing women in the Waikato, such as Waterford Birth Centre, River Ridge East Birth Centre and Waikato Hospital.

Additionally, over the past decade I have had numerous conversations, formal interviews and focus groups about birth and the spaces of birth with pregnant women. While in this chapter I do not draw explicitly on these data I mention them because they have helped shape my thinking on the spatiality of birthing.

My thinking on this topic has been shaped also by my own experiences of birthing. I have two children (born 1991 and 1996). In both instances I planned a home birth but on my midwife's advice' went to Waikato Hospital part-way through the labour because I became dehydrated and labour was progressing slowly. In the final instance, however, both hospital births were 'natural' and I returned home several hours after delivery.

COLONIZING AND RECLAIMING BIRTHING SPACE FOR MĀORI WOMEN

Angeline has given birth in hospital and at home. Her experience is in no way meant to represent Māori women more generally. It is difficult, and not always useful, to generalise about Māori women's experiences and understandings of hapūtanga (pregnancy) since they are a diverse group occupying varying socio-economic positions within specific hapū (sub-tribes) and iwi (tribes).[2]

Having said this, there are perhaps some general trends amongst Māori that are worth noting. First, Māori women are likely to have more pregnancies than non-Māori women. 'The fertility rates for the December 2003 year indicate that, on average, New Zealand women are giving birth to 1.95 children' (Statistics New Zealand 2004: 31, available <www.stats.govt.nz >, accessed 16 March 2005). 'Fertility rates for the December 2003 year indicate that the 30–34 year age group (114 births per 1000 women) is the most common age group for child bearing, only slightly ahead of the 25–29 year age group (109 births per 1000 women)' (ibid: 32). In 2003 the fertility rate for Māori women was 2.55 births per woman compared with 2.18 for all women in New Zealand (ibid: 33). This is despite the fact that there had been a rapid decline of Māori fertility in the 1970s which coincided with the equally rapid urbanisation of the Māori population.

The second trend worth noting is that Māori women are likely to become pregnant younger than non-Māori women. For Māori women in 2003, 45 percent were aged under 25 years (ibid.). 'In the December 2003 year, the median age of Māori mothers was 26.1 years, four years younger than for the total population (30.2 years)' (ibid.). In comparison, for the total population there has been a shift in peak child bearing years from the 20–24 age group to the 25–29 age group and very little decline in child bearing at older ages since 1971. This difference in age patterns of child bearing suggests non-Māori women limit their fertility by deferring the onset of child bearing to older ages, whereas Māori women are more likely to have their children early and then end their child bearing years at a younger age. The tendency for Māori women to have children at earlier ages than women in the total population is also reflected in the teenage fertility rate (Statistics New Zealand 2004, available <www.stats.govt.nz>, accessed 16 March 2005).[3,4]

Between 1971 and 1991 Angeline gave birth to seven children — four in a hospital and three at home. She has also been present at the birth of a number of her grandchildren. Well-known midwife and home birth advocate Joan Donley[5] assisted Angeline at home with the birth of her fifth and sixth babies. Another well-known midwife and home birth advocate Maggie Banks assisted Angeline at home with the birth of her seventh baby (see Ogonowska-Coates 2004 and Papps and Olssen 1997 on midwifery in New Zealand).

Angeline explains that since having her first baby in 1971 she has seen many 'changes of control, technology, the movement from a technologically driven process into something where we're allowed to be normal — natural ... I was 22 when I had my first and 42 when I had my last.' Angeline grew up hearing stories about her grandmother being assisted by her grandfather.

> Ordinarily they talk about bodies being tapu, women are tapu and men are hands off. In some cases, the birthing woman would kneel and lean forward grasping a pole for support during labour. Or, according to my mother, when her mother was giving birth, she would lean on my grandfather. My grandfather would sit facing her. She would kneel between his knees and use them as support to bear down as she delivered her child. He had an active role to play, but things have developed over time. Colonial influences came in with attitudes that having babies was a woman's job.

Angeline says that women's birthing bodies are generally regarded as tapu. This is still the case today (see August 2004 and 2005 on Māori women's menstruating, pregnant and birthing bodies and spaces as tapu). Something is in a tapu state when it is under the influence of the atua (atua meaning god or spirit). Its opposite is referred to as noa (not under the influence of the gods or spirits) (see Hanson 1982: 344).[6]

Historically a whare kohanga ('nest house' — Best 1975) or hut was specially constructed in order to accommodate Māori women just before labour, during labour and after the birth (Best 1924, 1975; Makereti 1986; Riley 1994; Tregear 1926). The hut was not erected in the village, but at some distance from it, and all such places were located where there would be no likelihood of crops being planted in the future. Only her birth attendants would be present with the pregnant/birthing woman.[7]

But colonization changed birthing practices for Māori. Angeline explains:

> In my grandmother's time what changed was that legislation came in and outlawed birthing at home. My grandmother having birthed all her children at home went to hospital for her last birth. The experience almost killed her. Instead of squatting she was made to lie flat on her back, to lose all her dignity. Her husband wasn't allowed to be around. According to my Mum, the last baby born, my aunt nearly died. Basically Māori women went, they went to hospital because they were forced to by law. There was this fear that you would go to jail or something would happen if you didn't do it. So then these maternity homes came into communities.

In the maternity wards Māori women were expected to birth 'in the Pakeha [European] way' (Donley 1986: 125; also on articulations of gender, race, ethnicity and class in colonial New Zealand more generally see Larner and Spoonley 1995). After the law changes Māori representatives in Taranaki petitioned the Committee of Inquiry for a Māori maternity hospital in Waitara — a facility that could provide first-rate care and retain Māori customs. 'Instead the Committee recommended special Maori wards be established in the public maternity hospitals, that is, Maori mothers would learn to have their babies in the Pakeha way' (Donley 1986: 125). Angeline comments that Māori women were expected to 'birth like Pākehā women so there was that taking away of control, that whanau [family] way of giving birth on your own land.'

As was typical in the 1940s and 1950s Angeline's mother, Tuaiwa (Eva) Richard,[8] gave birth not at home but in a hospital.

> In my mother's time she was totally brain-washed into that way of thinking. She had nine children and I can remember her picking up her bag with her big swollen stomach and having to walk about half a mile down a metal road to meet the taxi which would then take her to the hospital to have her baby and then in about a week or two weeks she was allowed to come home with this new baby. We would be looked after by Dad, aunts, cousins or our grandmother ... my mother's experiences were you went to hospital to have babies. That was it ... she had all nine in hospitals.

Donley (1986) argues that for Māori women, childbirth was, and continues to be, almost completely institutionalized within the Pākehā or European health system.

> It took less than twenty years for Maori childbirth to become institutionalised in hospitals. In 1937, when the recommendations of the Committee of Inquiry into Maternity Services laid the groundwork for the current system of maternity care, 83 percent of Maori women were confined at home, many with Maori lay midwives assisting at the birth. Nevertheless, the Committee recommended 'the same general policy of hospitalisation for Maori as for Pakeha women.' The provision of free maternity care under the Social Welfare Act 1938, assisted by the urbanisation of Maori women, achieved this by the 1960s, when in 1962 95 percent of Maori births were taking place in hospital. (Donley 1986: 122)

Donley (1986) explains that Māori women held out against having their babies in hospitals for as long as possible mainly because of tradition. It was customary for Māori women to birth at home with the support of whanau (family). Waiata were sung and after the birth the whenua (placenta) was returned to the whenua (land).

Maggie Banks (2000: 54) says 'In 1936 Maori women died in childbirth at twice the rate of European women. Their infants died at nearly four times the rate of their European counterparts.' The 1937 Committee of Inquiry reported that home births at this time were often taking place in overcrowded, impoverished and unhygienic conditions. Banks (2000) explains that today this high death rate is often used to make the point that traditional Māori practices were not safe. What it in fact indicates, according to Banks, is that Māori were suffering from the effects of colonization — tribes were forced from their land, demoralized and impoverished. Banks (2000) points out that 'Rather than addressing the issues of colonization, which created the change from robust health, declining health for Māori women in childbirth was gradually subjected to the same controls as for European women'.

One form of 'control' that Angeline points to is routine ultrasound scans of pregnant women's uteruses. Angeline was pressured by an obstetrician in the early 1980s in Auckland to have a scan when pregnant with her third child.

> She [an obstetrician] said 'right, you're off for a scan' and I said 'Excuse me, what is a scan?' because this was new technology and she turned on me and said 'are you a feminist?' I didn't even know what the word meant at that time. I said 'I'm about to be a mother again and I don't know what you are talking about. I want to know what a scan is'. And she said 'you feminists are all the same' ... I refused to have a scan and I went back home and wrote a letter to the Minister of Health at the time — Michael Bassett — and I said 'there are real issues that I have about Waitakere Hospital and how women are treated.'

This incident angered Angeline to the point that she sought a way of having a baby that took her outside 'mainstream' birth clinics and hospitals. A friend advised her to get in touch with home birth midwife Joan Donley. Angeline explains:

> I said Joan 'Look I'm just so frustrated, I'm 36 weeks pregnant, this baby is due soon and I have to go through this stupid process with this woman [obstetrician]. She said 'well you don't have to.' She said 'if I can find a GP [General Practitioner] and I think I can find one that can cover your pregnancy then I'll help you deliver it, I'll help you' ... this was my fifth child in 1982. So I met Joan for the first time. She told me a little bit about herself — that she'd been delivering babies back in Canada. This Pākehā woman with [her hair in] a bun — she said 'I'll come to your place,' so she came. I was really comfortable with her. So I said 'that's it, I'm not going to hospital. I'll have it at home.'

Angeline's decision to have a home birth was not met with universal approval. Several of her neighbours in Crystal Avenue, Glendene [Auckland], said things like 'oh she's mad' and 'something is going to happen.' In the weeks leading up to the birth Angeline felt quite 'isolated' from her neighbours apart from Maggie next door. Her mother also had concerns despite Angeline's reassurances that it was 'not a problem.'

Angeline continues her story:

> So anyway this baby decided to come two weeks early so I only had two weeks that I'd known Joan ... This baby arrived pretty quickly but Joan was there, the kids were all around, it was very comfortable, you didn't have to rush anywhere. You just sat and you had the baby basically where it was produced, you know, in your own bedroom.

After the baby was born the neighbours were quick to rekindle their relationship with Angeline and many came knocking on the door wanting to see the new addition to her family and keen to share their own stories about being born at home (see Figure 6.1).

For Angeline and her partner Alan home birth was about personal choice but it was also a political decision to reclaim control of the birthing process for Māori.

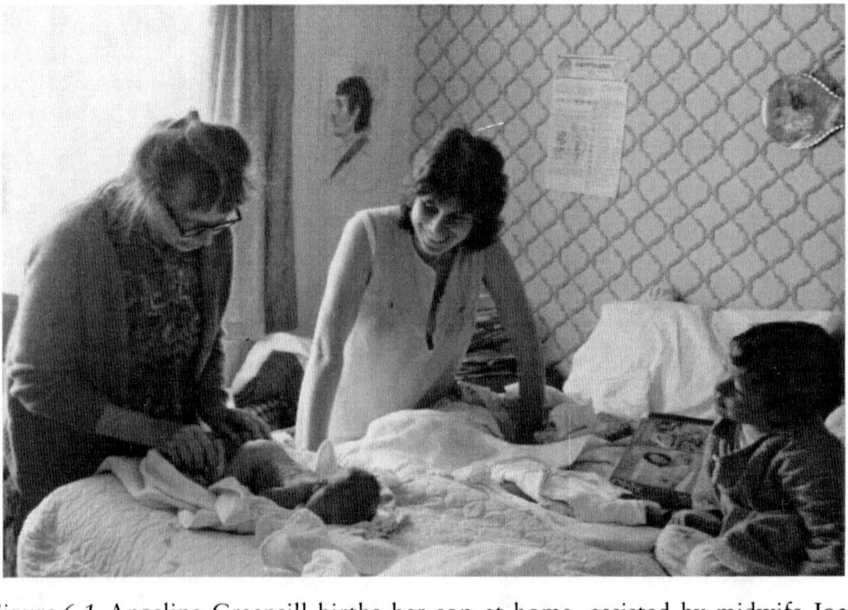

Figure 6.1 Angeline Greensill births her son at home, assisted by midwife Joan Donley, 12 November 1982. *Source:* Photograph used with permission of Angeline Greensill.

> I think I was probably one of the first Māori mothers to reclaim the power to birth at home, like back in Raglan. Our own people then started to take control over what they had done normally, done naturally. Alan and I talked about it and said 'look, children live and die all the time, babies are born, babies die, it doesn't matter where they are born. It could be in a hospital but they are covered up — you never hear the real statistics until you open up some of the comparisons in the home birth manuals and you see that babies die everywhere.' So for us we took this step knowing full well that if anything happened people would say 'well you should have done this, you should have done that.' We thought 'are we strong enough to do it?' I felt quite confident given that I'd had children before and knew it wasn't going to be a problem. I ate well, I looked after my health, and if you've got someone who supports you then it's okay.

Angeline also birthed her next child at home assisted by Joan Donley. Her last baby was born at home in Raglan (where Angeline comes from). For this birth Angeline required a Waikato-based home birth midwife so Donley suggested Maggie Banks: 'She [Banks] came to Raglan ... she was excellent. She was so laid-back and I was really comfortable with her.'

One of the things that Angeline really values about being able to birth at home is that family are able to be part of a significant event. She explains:

> When I was actually birthing in my room, my children were part of that event. I mean I've got five boys. They know where babies come from. They've been part of the process. I do it very discreetly on my knees, or standing, it's a position to deliver in and all they see is the baby coming out so you know, they've participated. They know about that stuff.

Angeline's mother, having birthed her own children in an era when hospitalization was the norm, was not as comfortable as Angeline's children with home birth. Angeline says:

> My mother stayed in the room next door doing a karakia [prayer]. She didn't witness the birth but she came in once the baby was born ... I was trying to get her to work through her fear, to remove images that she had of her own birth experiences by actually including her in a different sort of experience because for me it's always been really, well not enjoyable but empowering.

For Angeline home represents a space of escape from the hegemony of Western medical approaches to birthing. It also represents a space in which her whanau (family) are able to be part of the birthing process. Donley explains that Māori women resisted having their babies in hospitals for many years because according to Māori custom

> birth was a natural event that took place at home with support of the whanau. As one woman who had all seven of her babies at home has related, 'I was scared of hospital and had all my babies at home. Having babies then was just like a picnic. Members of the whanau were present and they would sing waiata oriori during labour and when baby was born they chose a suitable name from the waiata.' (Donley 1986: 122)

Angeline also conveys a strong sense of birth as a whanau event and in recent years has been involved in the birth of some of her grandchildren. She explains:

> I have seven grandchildren so every time the girls go into labour they ring me up — the boys ring me up and say 'Mum, we're ready to birth' ... my daughter rang up and said 'can we come and birth at your place?' and so the whole family came together over at our place because we've got a bigger lounge. She got on with her job and the birth was a very quick birth. Everyone said 'Let's have a cup of tea' ... And when she had her second baby at home she thought this is going to be a piece of cake, it was a whole night of labour. It was really hard labour but that was sort of birthing stuff, but [the] sort of environment I think is a very comforting environment for people, familiar with the place, you can feed, you can just carry on a normal way of life and it just becomes a normal part of life — just life really. She turned up this week at 9.00am and had a healthy baby boy at 10.45am assisted by Heather, a Māori midwife who has been at all three births.

Angeline's narratives about birthing at home remind me of bell hooks' (1991: 33) narratives about home. hooks argues that home in the face of racist oppression is a refuge, a space of 'care and nurturance', for black people. Home has a 'radical political dimension' (ibid). It is a site of resistance against racial apartheid. It is the domain of black women. hooks explains that when she was a girl she traveled across town — 'passing that terrifying whiteness' (ibid) to her grandmother's house. She says:

> I speak of this journey as leading to my grandmother's house, even though our grandfather lived there too. In our young minds houses belonged to women, were their special domain, not as property, but as places where all that truly mattered in life took place — the warmth and comfort of shelter, the feeding of our bodies, the nurturing of our souls. (hooks 1991: 33)

Both Angeline's and bell hooks' narratives about home illustrate that a complex (post)colonial and gender politics inhabit the space of home. Angeline captures this when she says 'They [home birth midwives Donley and Banks] are crossing the boundaries into a Māori way too, you know, I mean it didn't matter if you were Māori, Pākehā, whatever, the women went to where the women were.'

FEMINIST POLITICS OF HOME BIRTH

Since August 1990 when the Nurses Amendment Act was passed women in Aotearoa have enjoyed greater choice in terms of 'care' during pregnancy and birth. Before the Act was passed, doctors were legally in charge of all births. Midwives attended all mothers during labour, but they were required either to stand aside as the doctor took over just before the birth, or to deliver babies under a doctor's supervision (McLoughlin 1993: 56; see also Consumers' Institute 1994). For more than 15 years it has been possible for women to birth at home under the care of a midwife only and yet there is still widespread resistance to this. Any incident of home birth that 'goes wrong' tends to be fodder for the media. Accusations about irresponsible mothers, parents, midwives and sometimes other health professionals tend to fly (Banks 2000). It is not surprising, therefore, that organizations such as Home Birth Aotearoa (a national organization that supports and celebrates home birth) and home birth midwives sometimes counter this resistance by (over) 'selling' the notion of home as an idealized site — the best possible place to birth.

In short, those in support of home birth tend to (re)present homes as places that are essentially 'good', nurturing, supportive, 'natural', 'normal' and feminized. Homes are understood to be spaces in which women can reclaim control over their birthing bodies. Hospitals, however, tend to be (re)presented as places that are 'bad', neglectful, unsupportive, unnatural, abnormal, and masculinised. They are places in which men control women's birthing bodies. Donley comments (back cover Banks, 2000) 'In moving from "Her Home to His Hospital" — from 'nurture' to 'management' — women have been bereft of their knowledge and skills, that is, power over normal birth.' Donley was one of the founding members of the Home Birth Association in New Zealand and over the past few decades has published several books examining the politics of home versus hospital births in New Zealand (see Donley 1986 and 1992).

Over the past few decades there has emerged a dichotomy between medicalised births which are considered to be unnatural and home births but also water births, active births, births in quiet or tranquil darkened rooms, and gentle births which are considered to be natural. The natural birth movement emerged in the United States in the 1950s as a reform movement to do away with the excesses of medical management (Katz Rothman 1982). A number of the recent 'alternative' approaches to birthing are based on the idea that women ought to do what comes 'naturally' (that is not to say that women ought to be entirely unprepared for birth — there are classes on natural childbirth). Approaches to birth by people such as Michel Odent (1986), Janet Balaskas (1989) and Frederick Leboyer (1975) tend to be based on the notion that birth can be, and should be, a natural event. Kitzinger (1989: 170) in discussing Odent's (1986) approach to birth claims that: 'For him, the most important thing is to provide an environment that facilitates a spontaneous psychophysiological process in which the woman who is left undisturbed will feel as if "on another planet".' Kitzinger (1989: 170) says he 'has rediscovered the use of water in birth.' Balaskas (1989) has developed a method of preparing for birth based on yoga. Balaskas advises women in labour to move around and change positions frequently. She also advises women to give birth in a squatting or kneeling on all fours position. Kitzinger (1989: 317) claims that, according to Leboyer (1975), 'we must focus on a baby's needs and learn how to reduce its suffering' immediately after birth. In order for this to happen 'the birth room must be calm and hushed, the lights are dimmed and those handling the baby must do so slowly, carefully and lovingly. This is known as *gentle birth*' (Kitzinger 1989: 317 italics in original).

The discourse of natural childbirth has also carried with it a tendency to compare pregnant women with animals (usually mammals) which are frequently considered to exist outside of culture. Sometimes, more specifically, a correlation is made between birthing women and birthing animals. Kitzinger (1989: 192) advises birthing women to learn from mammals. She writes: 'If we watch any mammal giving birth, a cat, for example, or a sheep, we notice that she does not take great breaths in and then 'block' the birth canal by holding her breath.' Kitzinger (1989: 192) says: 'A sheep gives birth with rather light, quick breathing. Her breath is involuntarily held as she bears down and then she continues the light, accelerated breathing again.' Embedded in advice such as this is a notion that nature and instinct can be separated out from culture and social constraints. The main point that advocates of natural childbirth make (and natural birth is often associated with home birth) is that women would instinctively act like other mammals if medical science did not interfere.

The difficulty with this argument is that the pregnant human body can never be fully extracted from culture, it is always already socially and politically coded through a range of competing discourses. Maternal, reproductive and birthing bodies are as constructed as any other bodies. Like all bodies they are a product of nature and artifice. Much of the literature that advocates for home birth, however, treats women's birthing bodies as though they are natural and essential and treats homes as though they are always the locus of idyllic family life. Banks (2000: 5: italics added) explains that there are home birth midwives in New Zealand

> who decide they will *only provide services to women choosing to home birth,* unless there is a medical indication for the birth to occur in hospital. It is, therefore, only in the transfer situation that they will attend woman in hospital. This decision comes from a deep conviction that it is *only at home* that a woman can achieve her true autonomy, and experience birth as a family event, in a Wise Woman way.

Some home birth advocates, in an attempt to underscore the benefits of home birth, reify home as an idealized site. It is argued that birthing at home in a familiar environment (unless there is a medical reason for birthing in a hospital) is always heralded as the best option for mothers and babies. Home birth is unproblematically portrayed as a positive choice offering benefits such as safety, less intervention than at a hospital, no medication, less infection, familiar environment, no travel, fewer breastfeeding problems, continuity of midwifery care, privacy, emotional bonding with the baby, and the emotional well-being of the mother (see Banks 2000; Donley 1986).

This may well be the case (in many but not all instances) and it is clear that midwives have had to fight hard to offset some of the negative interventionist birthing experiences associated with hospitals (Donley 1986) and negative images of home birthing as unsafe and irresponsible. Angeline explains:

> I think you have to be very strong within yourself really, with everything that is out there in the community, to actually take those sort of steps and make those decisions [to home birth]. ... There's a resistance [to home birth] and it's building again. I mean the medical profession when I was having babies in the 1970s were totally in control and it was mainly males. You'd see mainly males coming through, very few female obstetricians ... it was mainly a male driven field and you dreaded going to some of those big hospitals because everything was so medicalized so I was really lucky to meet Joan Donley and get that other experience of what can happen when you just take control of your own life and you know that's a very empowering experience. The [home birth] movement's not huge and when you get media publicity that focuses on things that occur, children that die, which is really sad but they never focus on the hospital births ... and it's always been the same from way back, men have wanted to control and that comes through I think. So, I've always been very supportive of midwifery. I expect standards of course but you know.

Perhaps in an attempt to offset some of the negative perceptions of home birth feminist and postcolonial advocates have tended to make assumptions about a mother's home environment, such as that it is private, safe, clean, comfortable, and loving. This is not surprising given that as Johnston and Valentine (1995: 99) explain:

> Home is a word that positively drips with associations — according to various academic literature it's a private, secure location, a sanctuary, a locus of identity and a place where inhabitants can escape the disciplinary practices that regulate our bodies in everyday life.

Somerville (1992 cited in Johnston and Valentine 1995) picks out seven key dimensions of home: shelter, hearth (that is, emotional and physical well-being), heart (loving and caring social relations), privacy, roots (source of identity and meaningfulness), abode and paradise ('ideal home' as distinct from everyday life) but these meanings of home do not resonate with everyone (see Blunt and Dowling 2006).

For example, as Johnston and Valentine (1995) point out, home is often conveyed as being synonymous with the heterosexual 'family' and the ideal of family life and yet it can take on very different and contradictory meanings for sexual dissidents who share a house with heterosexual family members or who live in a variety of other household configurations (Johnston and Valentine 1995: 99).

Women's homes are not necessarily their castles. Not all women live with, or want to live with, a partner and/or other children. Women sometimes live alone, with relatives, with friends, or with 'flatmates' or boarders.[9] Their homes may not feel like the best place to birth. Homes can take on different meanings for women who are abused by 'loved ones' or acquaintances (see Johnson with Huggins and Jacobs 2000: 79–80 on 'the politics of male violence') or for women who live with others with whom they feel no connection, or for women whose houses are lacking in privacy, or are cold, noisy, and uncomfortable. In part, as a political strategy, home birth advocates have not tended to hold up the notion of home to this kind of critical scrutiny.

Many feminist geographers and others (e.g. Matrix 1984; Spain 1992) over the past few decades have, however, examined the politics and social meanings surrounding domestic space. Women's labour — sometimes birthing but more often cooking, cleaning, laundry, nurturing, gardening, and decorating — has been the subject of much discussion (Katz and Monk 1993). Sometimes women may want to birth in hospital because it represents an escape (no matter how temporary) from domestic labour at a time when it is most needed. Other women may want to birth in hospital for medical reasons such as they have diabetes or a heart condition. Some women report that their husbands or partners are anxious about or against home birth (see Shaw and Kitzinger 2005: 4–5 on some of the reasons women provided on choosing or not choosing a home birth.) It is too simplistic to say that women don't want a home birth simply because they have been duped by a colonial and/or patriarchal medical profession into thinking that hospital is the only choice. Whilst structural political, social and health arrangements cannot be ignored (i.e. it is not necessarily easy for women to exercise their right to choose the place they want to give birth), women in many countries including New Zealand and the United Kingdom (see Shaw and Kitzinger 2005) do have the right to decide to birth at home, regardless of whether they are expecting their first or subsequent children or of any perceived 'risk' factors. Such decisions are often complex. There is no neat and tidy divide that positions homes as necessarily 'good' and hospitals as 'bad' and yet debates about where and how to birth have tended to become constructed around such dualisms.

DECONSTRUCTING THE SPACES OF HOME AND HOSPITAL

Scott Sharpe (1999: 95) provides a set of dualisms that characterize the politics of childbirth:

feminine/masculine
natural (drug-free, non-interventionist)/medicalised (medicated, interventionist)
midwife/obstetrician (obstetric nurse)
home (personal)/hospital (institutional)
private/public
normal (healthy)/abnormal (potential complications)
technique/technology
risk minimization/risk elimination
autonomy/paternalism
the touch/the gaze
dynamic/confining
individual/standarised

Sharpe (1999: 102) explains 'Where the strong polarization of the debate into medicalised and non-medicalised birth has led to a certain restriction of practices, demassifying these poles may allow for a greater diversity of experiences to count for a "significant" or "successful" birth.' Sharpe argues that birth centres are spaces that have opened up possibilities for alternative practices (also see Fannin 2003). He focuses on a particular birth center — the Birth Centre at The Royal Hospital for Women (RHW), Sydney, Australia). This Centre, and others, argues Sharpe (1999: 93–94), destabilize the binaries listed above in that they signify both 'sites of resistance to patriarchal obstetrics' and sites of 'co-option which have arisen to counter the growing home-birth movement.' Sharpe adds that in the RHW Birth Centre's promotional literature represents the Centre as an intermediate space between home and hospital. Birth centres and the practices that occur within them are both constituted by, and constitutive of, the dualisms listed.

Fannin (2003: 513) says that relatively recently in the United States a number of hospitals have transformed their birthing wards into 'homelike' birth rooms 'with hardwood floors, brightly colored bedspreads, rocking chairs, and carefully designed cabinetry that masks medical equipment.' Some rooms at Waikato Women's Hospital in Hamilton, New Zealand, have also been transformed into more 'homelike' spaces. Fannin argues that the newly 'personalized' environments of hospitals are offered as evidence that the medical 'establishment' no longer views childbirth as illness but Fannin (2003: 514) suggests more is also going on. These new birth spaces can also be linked to 'neoliberal economic logics that blur the boundaries between market, home and hospital.'

Independent birthing centres that are not linked to existing hospitals also represent attempts to market services to new 'consumers' of birth. Waterford Birth Centre in Hamilton, New Zealand, is an interesting example that illustrates the changing spatiality of birth. The Centre is described on their website as a 'primary birth centre for low risk births, offering a peaceful, safe, friendly environment for women in the Waikato area to enjoy their birth experience and postnatal stay' (Welcome to Waterford Birth Centre, 2005, available http://www.waterfordbc.co.nz/, accessed 7 June 2005). The virtual tour of Waterford Birth Centre shows photographs of the entrance (which looks like a driveway to a home), the reception area, antenatal classroom, large private birthing room (which has adjoining spa bath and ensuite), single postnatal room, and double postnatal room (both rooms have yellow walls and curtains). Photographs of the postnatal rooms contain mothers, babies, small children and one image shows a man (presumably a father) sitting on a bed. The images are a mixture of signifiers of hospital and home. The website boasts 'Waterford has a quiet tranquil setting with views over the Waikato River. Natural light and access to river walks complement the modern amenities provided' (ibid). Services provided by midwives at Waterford include pregnancy tests, antenatal care, referrals to specialists, antenatal classes, midwives' rooms, labour and delivery care including water births, postnatal care and postnatal visits and referrals to child-care services.

A woman called 'Karen' posted the following comment about the Waterford Birth Centre on an Internet site:

> My niece's Mother gave birth in Waterford in Hamilton. Christy and I assumed it must be private because it is sooo [sic] flash but it is public! The only way to describe it is like your own hotel suite with a spa pool in each room! They also put a photo up of baby as soon as it is born so that people who are overseas can get a quick look — aww! I am going to look into the birthing centres when I get to that stage — I don't want it to feel like a sterile environment that I am not comfortable in and I certainly don't want to be sharing a room with three other babies! How much sleep would you honestly get?! (Better Living Discussion Forum, 2004, available <http://www.betterliving.com.nz/net/forums/>, accessed 3 December 2004)

Figure 6.2 River Ridge East Birth Centre, Hamilton, New Zealand. *Source:* Photographs taken by Robyn Longhurst, compiled by Betty-Ann Kamp.

Interestingly Karen perceived the facility at Waterford to be like a 'hotel suite' rather than a home but she does note with approval that it is not a sterile environment.

Another publicly funded birth facility in Hamilton that attempts to deconstruct the binary between hospital and home is River Ridge East Birth Centre (see Figure 6.2). Clearly both Waterford Birth Centre and River Ridge East Birth Centre are much smaller facilities than the large 600 bed base Waikato Hospital where approximately 3000 babies are born each year (see Figure 6.3). River Ridge East Birth Centre claims they are a primary birthing facility for women having low risk births. It is also possible for women wishing to have their baby at Waikato Hospital to transfer to the Centre after their baby has been born for postnatal care.

The River Ridge East Birth Centre opened on Christmas Day 2002. Their website states: 'Women who choose to stay at River Ridge East Birth Centre can enjoy a relaxing stay and get to know their baby in a homely environment that is both welcoming and comfortable.' (River Ridge East Birth Centre, 2006, available <http://www.riverridgeeastbc.co.nz/>, accessed 4 September 2006.

> The environment at River Ridge East is caring and welcoming for both women and their families. ... The four birth rooms are fully equipped, with ensuites and monitoring equipment. Many women use the birthing bath or pool for relaxation and pain relief. The postnatal rooms are warm and comfortable, with soft, soothing décor. Each room is

Figure 6.3 Waikato Hospital, Hamilton, New Zealand.
Source: Photograph published courtesy of Hamilton City Council (http://www.hamilton.co.nz).

> equipped with an ensuite and individual air-conditioning. A fridge and self-service area for tea and coffee making are provided in each room along with a TV and private phone. There is a large window seat where a support person can comfortably sleep. The postnatal rooms open onto courtyards with seating around a small water feature. (Maternity Information Booklet, River Ridge East Birth Centre, 2006, available <http://www.riverridgeeastbc.co.nz/index.htm>, accessed 18 April 2006)

The social and physical environment at the Centre is constructed in the promotional material as a space that allows for individual women and their families, for domesticity and technology. It is represented as offering the warmth and comforts of home and technological advantages of hospital — an in-between space.

CONCLUSION

This chapter has examined the importance of spatiality in relation to birth. In New Zealand birthing has moved from home to hospital. More recently, however, some women have chosen to return home to birth. Others chose to birth in birthing centres which attempt to combine elements of home and hospital. Postcolonial and feminist discourses have represented homes as essentially 'good' and as the best place to birth. Banks (2000: 147–148) says 'The sensual nature of birthing demands a birth scene that resembles the intimate environment in which the baby was conceived. It is this environment, which frees the woman from her need to be vigilant and ensures her maximum relaxation.' Birthing wards in hospitals, however, are represented as essentially 'bad' and as the worst place to birth.

Whilst it is understandable that these discourses of home as haven have emerged in response to often oppressive racist and masculinist structural relations that govern women's birthing bodies, nevertheless the notion of home needs to be problematized in such discourses. This is because neither bodies, nor places, are essential. Homes and hospitals do not have essences but are imbued with culture and history. It is important to consider carefully the politics of spatiality before drawing easy conclusions about home birth. The binary division between home and hospital that has characterized the politics of childbirth needs to be deconstructed. Birthing centres represent one way of troubling this binary.

Many postcolonial and feminist geographers (e.g. Johnson with Huggins and Jacobs 2000; Johnston and Valentine 1995; Watson 1988) over the past decade have illustrated that it is possible to 'read' social relations in architecture. It has become evident that the notion of home varies for different people. For some it is a space of renewal, a space to escape oppression and colonial practices. For example, for some Māori, including Angeline, the shift from lay midwives and tohunga (spiritual leaders who were fundamental to the maintenance of Māori health) to registered midwives and doctors, from birthing with whanau (family) to birthing with strangers was oppressive. Having the option to (again) birth at home offered a sense of liberation. For others, however, home may be a site of subjugation. These more contradictory and complex readings of home need to be kept in mind when considering discourses on home birth.

Homes reflect and reinforce social meanings (Spain 1992; Weisman 1992). They are imbued with culture and history which is not fixed or unitary. This makes it possible, therefore, for women to experience birth as a family event in a labour ward or to experience alienation and oppression at home. This statement is not meant to deny the very 'real' (masculinist and racist) structural arrangements that are constituted within these spaces (hospitals and homes) but multiple scenarios in both spaces are possible. Some hospitals are undoubtedly highly eurocentric, masculinist spaces but so too are some homes.

This chapter is an attempt to trouble the spaces of homes and hospitals and to examine the potential of the discursive and material spaces in-between. Women perform their gendered identities differently in different spaces. For some women, in some instances, home can be a sanctuary. In other instances it can be site of oppression. Unproblematically coupling home with loving, caring, and sharing social relations is fraught. Postcolonial and feminist theorists and practitioners who argue for home as the one and only place where is it possible to have a 'successful' and 'fulfilling' birth would do well to consider the politics of gender and domestic space more critically. As Spain (1992: 62) points out: 'Whether a woman gives birth at home, in a birth center, a hospital birthing room, or the hospital's labor/delivery unit, the choice should be hers.' Providing women with an understanding that the spatial form of private and public architecture — homes and hospitals — both reflects and reinforces institutionalized racism, sexism, ablism (see Chapter 4), classism and so on will help equip them with the knowledge necessary to make informed decisions about birth.

7 'Queer Breastfeeding'
(Im)proper Spaces of Lactation

Australian feminist Fiona Giles (2004: 301) examines 'breastfeeding behaviours [that sit] outside the normative constraints that apply in contemporary Western culture.' She refers to these behaviours as 'queer breasting'. Giles (2004: 301) explains the 'contemporary norm is that between 50–90 percent of mothers in industrialized nations will begin breastfeeding their babies before leaving hospital, and wean them within the first year.' Some breastfeeding, however, sits outside this norm. In her book *Fresh Milk: The Secret Life of Breasts*, Giles explains:

> Western culture tends to think of women's milk only in relation to the babies they might choose to feed. We think of lactating women as immediately postpartum, still wrapped in the aura of childbirth and laden with the trappings of infancy: bassinets, strollers, blankets, diaper bags ... lactating breasts are connected to the hospital bed, to nursing bras and nursing pads, and special diets, and sleepless nights, and hollow-eyed jealous husbands. They are medically scrutinized, gently massaged, and intensely discussed during the time they are used. Then, when the flow is stanched, they are primly returned to their lacy cage, to be admired again according to purely aesthetic considerations. (Giles 2003: xii)

The aim of Giles's book is to examine the social meanings of breastfeeding (as opposed to the nutritional content or medical benefits of 'mothers' milk') which have not to date been paid much academic attention. Giles covers subjects such as cooking with breast milk, donating milk to a milk bank after the death of a child, adult nursing, and lactation pornography as well as some of the more rehearsed subjects such as weaning an older child and mastitis. Well-known popularist writer on pregnancy and birth, Kitzinger, claims that Giles's book says 'All the things that the other books about breastfeeding don't say' (front cover).

As may already be evident from this introduction, Giles's work on 'queer breastfeeding' largely prompted this chapter. The chapter in a sense resonates with Chapter 2 on women coming out as pregnant at work. Some breastfeeding behaviours, like pregnancy at work, remain closeted. Giles's work on 'queer breastfeeding' enabled me to begin to think through some of the spaces — material, discursive, and imaginary — of breastfeeding. Lactation has long been a symbol of maternity — sustenance, warmth, goodness — and yet there is a vast and diverse range of lived experiences of lactation, some of which reiterate dominant discourses of the 'purity of mothers' milk' and others that contest these discourses. The chapter discusses what might be considered some of the more unusual experiences of lactation in order to understand more about the spatiality of lactation — where and when lactating breasts are understood to be in place and out of place.

The chapter puts forward a number of examples of 'queer breastfeeding'. First, it examines 'adult nursing' focusing on the case of a prostitute offering breast milk as an optional extra to clients in a brothel in a small New Zealand town. Second, is a discussion of cross-species nursing focusing on another recent story in the New Zealand press that relays the story of a woman who breast-fed her Staffordshire Bull Terrier puppy. Third, is a discussion of the phenomenon of male lactation. Fourth, another story from the New Zealand media is discussed, this time the story of a mother who still breastfeeds her daughter who is nearly five years old. The chapter concludes with an examination of why 'queer breastfeeding' prompts so much cultural anxiety and links this to issues of spatiality.

ADULT NURSING

'Brooke', a 25 year old prostitute in Harewa, a small town in South Taranaki, New Zealand, attracted media attention when she placed an advertisement in a local publication offering 'the milk of human kindness' to her clients. Brooke, who gave birth six weeks earlier to a son was 'offering customers fresh breast milk as an extra service' (Wetton 2003: 3). Brooke and parlour owner Ruth Bain claim business has picked up since they began running the advertisement (Wetton 2003: 3). A current affairs television programme (*Holmes* 18 July 2003) featured Brooke explaining that she did not breastfeed her clients (although they could ask to taste her breast milk at an extra cost) rather it was 'the idea of milk being in there' (quote from interview with Brooke on *Holmes* 18 July 2003) that was supposedly arousing.

The advertisements were the talk of Hawera. 'Members of the public questioned in the streets ... were mostly queasy about the activity and its brazen advertising' (Wetton 2003: 3). Concerns were raised on behalf of the baby by the breastfeeding advocacy group, La Leche League. Director of the League, Rosemary Gordon said 'There was a risk infectious diseases such as herpes, hepatitis and tuberculosis could spread between the baby, clients and the mother' (*Waikato Times* 18 July 2003: 3). Gordon said that it would be a concern if the baby's milk supply was being compromised. Journalist Krysti Wetton reports in *The Daily News*:

> Brooke said she washed herself with hot salt water and showered before feeding her son, and did not believe there was a risk of cross-infection. 'My kid comes first,' she insisted. She started offering her breast milk after a suggestion from a client. 'Most of these guys are clean cut, well-presented nice guys,' she said. They were generally aged over 30 and one man, who lived in Wellington, told her he would be traveling to Hawera for a repeat serving. (Wetton 2003: 3)

The health of the baby, however, was not the only issue at stake. It did not take long for moral outrage to surface. The television reporter on the current affairs programme *Holmes* asked Brooke whether she considered offering her breast milk to clients while breastfeeding a young baby to be morally 'pure'. *The Daily News* asked people on the streets of Hawera what they thought of the offer. One woman who had recently stopped breast-feeding her child commented 'I wouldn't want men hanging off my breasts having a feed. It's a bit sick isn't it?' Another Hawera resident, Jason, said he had tasted his partner's breast milk, 'But what you do in your own home is up to you. This is nasty.' (Wetton 2003: 3)

In short, the act of breastfeeding an adult was represented as a potential public health issue and as immoral. Brooke's breasts are widely considered to be 'in place' at home (private space) with her baby, but 'out of place' at the brothel (public space). Her milk is seen by most people to be appropriate for her baby to consume but not for her clients. To date a number of studies have been carried out on mothers' experiences of breastfeeding in public and private places (e.g. see Mahon-Daly and Andrews 2002) but there is limited research on breastfeeding that sits outside of these mother–baby relationships.

An interesting epilogue to the story of Brooke occurred a month later in August 2003 when the New Zealand Ministry of Health changed its mind over using a poster at the start of a national breastfeeding campaign (Garner 2003: A3). The poster, A4 size, was of actor Michael Hurst at his office desk, with his shirt unbuttoned and a baby at his 'breast'.[1] It was meant to be distributed to 600 business owners and employers as part of an information pack on how to support breastfeeding in the workplace. A focus group of business owners and employers viewed the poster before its intended release reported that the poster captured their attention because it depicted a man rather than a woman breastfeeding. 'Had it been a woman they would have looked at it and decided it was a women's issue, and decided it was not something they had to engage with' (Garner 2003: A3).

It was suggested in the media that the Ministry of Health's decision to not distribute the poster was connected to the publicity a month earlier about Brooke offering breast milk to clients. Steve Chadwick, who spoke on behalf of the Health Minister Annette King said 'I understand their feeling that there has been enough negative publicity around breastfeeding when breastfeeding is meant to be a positive and affirming statement' (Garner 2003: A3).

In August 2001 actor Lucy Lawless (well-known for her role as Xena Warrior Princess) was featured on a poster for World Breastfeeding Week (WBF is an outreach vehicle for the breastfeeding movement which is celebrated in over 120 countries; see Lockett 2001). Lawless was breastfeeding her son while perched on the side of a chair. She is wearing fishnet stockings, a white shirt and a short skirt. The photograph is a mixture of the traditional and contemporary, Madonna and whore. The tagline below the photograph reads 'Breastfeeding — my best role ever' (to see the photo visit Women's Health Update Magazine, 2001, available <http://www.lucylawless.info/articles/womenshealth/>, accessed 5 September 2006). The poster was distributed to secondary schools, child-care centres, community groups, maternity providers, District Health Boards and other health professionals. 'While the response was largely positive, it provoked some controversy. A few people questioned the combination of breastfeeding and sexy sophistication' (ibid). If Lawless crossed a 'boundary' for some, then Brooke most certainly crossed a boundary.

Cindy Stearns (1999) argues that for women regarded as 'good maternal bodies' the prominence of the sexualized breast poses a problem.

> The good maternal body is not commonly believed to be simultaneously sexual, despite the obvious facts of human reproduction (Davis-Floyd 1992; Newton 1977). The sexual aspects of women and the maternal aspects of women are expected to be independent of each other. Thus, breastfeeding raises questions about the appropriate uses of women's bodies for sexual or nurturing purposes. (Stearns 1999: 308)

Stearns (1999: 308) notes that increasingly women are faced with the quandary of the sexual *or* the nurturing maternal breast. It is, therefore, little wonder that Brooke's offer of breast milk to clients caused cultural anxiety and moral panic. Young (1990: 132–33) explains breastfeeding and '[b]reasts are a scandal because they shatter the border between motherhood and sexuality.' Brooke publicly shattered this border and resisted cultural expectations to be a modest, 'good' maternal subject who uses her body — her lactating breasts — in a manner that is 'appropriate'.

It is possible to find a number of other examples of adult nursing (breastfeeding). Giles (2003: 97) explains over the past few years a number of Internet sites have been set up for couples who are interested in adult nursing. There is the 'Society for Nursing Couples' and 'The Community of Adult Nursing' (C.A.N.). Both these organizations offer detailed instructions on how to induce lactation and what to expect from a breastfeeding partnership. Members share stories and sometimes photographs. Giles

(2003) notes that although there is a sexual flavor running through many of the narratives, what is most remarkable are the references to the deep intimacy, closeness, and friendship that adult nursing partnerships seem to create. 'The overall tone of these websites is not unlike that of a politely earnest birth-watching community. Though explicitly detailed, their function is not necessarily sexual arousal' (Giles 2003: 97). C.A.N. disassociates their members from those who have a 'lactation fetish' which they claim is a fetish where 'the woman uses her milk as a sexual toy or lubricant. In some cases one or both partners will drink the woman's milk, but rarely is any real nursing involved' (Two hearts, one soul, 2003, available <http://www.twoheartsonesoul.com/lactation.htm>, accessed 2 May 2003).

When Giles (2003) put out a call for people to complete her questionnaire on lactation three men responded 'confessing' having fantasized about 'breastfeeding sex' for many years. Several other men contacted Giles by email (anonymously) seeking information about contacting lactating women. Engaging with Giles' research enabled me to understand Brooke and the media furor that surrounded her a bit differently. I realized that adult nursing is perhaps not that uncommon. It is, however, still deeply closeted. There are powerful cultural protocols surrounding women's lactating breasts. They are seen to belong in some (rather limited) spaces but not in others. In the West they are certainly not seen to belong in a shared space with animals.

INTERSPECIES BREASTFEEDING

In 2004 I came across a press clipping about a woman, Kura Tumanako, in Hastings, New Zealand, who was breastfeeding her Bull Terrier puppy. Tumanako's rationale was that she 'wanted the dog to grow up to protect her baby daughter' (Hodge 20 November 2004). 'Her baby had stopped being breast-fed and Ms. Tumanako did not want her milk to go to waste' (ibid). She said she did not care what people thought about her breastfeeding the pup: 'It's my life, my responsibility. I make my own choices ... I'm going to look after me, my baby and my puppy' (*New Zealand Herald* 18 November 2004: A5) (see Figure 7.1).

Animal behaviour experts raised concerns saying that 'the woman could end up with a savage dog if it was trained as a protector' (Hodge 18 November 2004). Within days of the story breaking it was reported that Tumanako had been admitted to a psychiatric unit and that her three month old baby had been removed from her care (Hodge 20 November 2004). No other information was provided because of confidentiality issues. Also cartoonists were quick to pick up and comment on the story of a woman breastfeeding a puppy (see Figure 7.2 and Figure 7.3).

It seems though that interspecies breastfeeding is more common than may first be expected. Giles (2003: xii) explains 'I came across a press

Figure 7.1 Kura Tumanako breastfeeds her pup, Honey Boy. *Source*: *New Zealand Herald* 18 November 2004.

clipping from March 1997, about a South African mother, Dristen Grobler, who had breast-fed a Poodle puppy while also breastfeeding her baby.' Human to animal cross-species breastfeeding raises anxiety, and yet we regularly drink the milk of cows without much thought or feelings of abjection. Giles (2003: 30) notes:

> Some women even expressed a willingness to feed a baby of another species, if it were in need. ... An adventurous lactation consultant wrote 'I have also nursed a baby puppy ... I wanted to feel what it was like. It felt like tugging, and almost tickled. I was amazed he continued to suck from my breast even though there was no milk, that I could see anyway.'

An Internet search revealed several stories of women breastfeeding or suckling pigs in the highlands of Papua New Guinea. Photographs included

Figure 7.2 A large litter. *Source*: *Waikato Times* 19 November 2004. Cartoon appears courtesy of Allan Hawkey and the *Waikato Times*.

a mother breastfeeding a small orphaned piglet and a woman suckling a larger pig (Papua New Guinea, 2003, available <http://www.uq.net.au/~zzlhiess/papua_new_guinea.htm>, accessed 3 May 2003). So, while cross-species breastfeeding may seem 'queer' to many living in the West, it is not 'queer' to those living in the highlands of Papua New Guinea. It seems that once again the spatiality of lactation is important. It makes a difference where breasts are understood to be in place and out of place.[2]

'MILKMEN': MEN WHO BREASTFEED INFANTS

Men's breasts, in an infant's mouth, are also considered to be out of place. In the first instance, men are not expected to have 'breasts', they are expected to have 'chests'. But men do have breasts (see Longhurst 2005b on 'man-breasts' as a material, discursive and psychoanalytic space of masculine embodied subjectivity). In the second instance, even if men do have breasts it's widely thought that they should not attempt to feed an infant with them. Most people do not know in fact that it is possible for men to lactate. The well-known website Wikipedia has this to say about male lactation.

Figure 7.3 Breast-fed as a pup. *Source*: The *New Zealand Herald* 20 November 2004. Cartoon appears courtesy of Rod Emmerson and the *New Zealand Herald*.

> The phenomenon of male lactation in humans has become more common in recent years due to the use of medications that stimulate a human male's mammary glands. Though human males have nipples, it is not so often understood that they also have mammary glands. Ordinarily the mammary tissue is low in volume and cannot be noticed. Under the appropriate hormonal stimulus — the hormonal stimulus that nature provides to human females when they become pregnant and give birth — the mammary glands of human males can also produce milk. The volume of milk produced will be small relative to the amount that a female can produce. (Wikipedia, 2005, <http://en.wikipedia.org/wiki/Male_lactation>, accessed 30 May 2005)

There are a number of Internet sites on lactating men, or 'milkmen' as one site refers to them (Bornfree, 2005, available <www.unassistedchildbirth.com>, accessed 3 June 2005).[3]

Consider the following newspaper articles. The first is entitled 'Widower learns to breastfeed.' The article explains:

> A Sri Lankan widower has attracted the attention of doctors for his ability to breastfeed his young daughter. Mr. B. Wijeratne, from Walapanee, near Colombo, took to breastfeeding her soon after his wife died three months ago while giving birth to their second child. His elder daughter, 18-month-old Nisansala Madhushani, was so used to her mother's milk that she would not take formula milk. Mr. Wijeratne told Sinhalese language newspaper Lankadeepa: 'My child would reject the powdered milk I tried feeding through a bottle. Unable to see her cry I offered my breast. That's when I discovered that I could breastfeed her.' The 38-year-old's ability to produce milk was noticed by doctors at the government hospital in the town of Kurunegala. Dr Kamal Jayasinghe, a spokesman for the hospital, said: 'Men with a hyperactive prolactine hormone can produce breast milk.' His younger daughter, who has still to be named, has taken to powdered milk. (Ananova, 2005, <http://www.ananova.com/news/story/sm_700634.html>, accessed 30 May 2005)

David Shanley, an American father who co-owns Bornfree, an organisation dedicated to unassisted childbirth, claims he was able to produce breast milk after the birth of his first child (Bornfree, 2005, available <www.unassistedchildbirth.com>, accessed 3 June 2005). While Shanley did not end up breastfeeding his child (his wife Laura breast-fed their child), the Shanleys wanted to prove that male lactation is possible.[4] On the website Laura Shanley also discusses a letter she received from a friend of hers about two gay men adopting a child. One of the fathers began using a breast pump, and by the time the baby was 12 weeks old, he was producing a full milk supply (ibid).

Despite this wealth of information on male lactation on the Internet, the topic has been ignored by academics in masculinity studies, feminist/gender studies, and certainly in geography. Research (contemporary and historical) has been conducted on women's breasts in relation to art, self-image, surgery, breast cancer, sexuality, fetishes, nursing and maternity but not on men's breasts.

Breasts are usually seen as a symbol of feminine sexuality (Young 1990: 190) and lactating breasts as a symbol of maternity. Men's breasted, and especially lactating breasted, bodies, therefore, tend to make little sense. They cause considerable cultural anxiety because they disrupt understandings of a specified kind of feminine maternity.

Male lactation is yet another example of what Giles terms 'queer breastfeeding'. It illustrates that breastfeeding needs to be understood as a 'cultural practice, rather than as something natural, innate, or even gendered.' (Bartlett 2004: 341)

'BREAST IS BEST FOR MATILDA'

Breastfeeding, it seems, must adhere to prescribed cultural norms. One of those norms is that children should not breastfeed for 'too long'. What constitutes 'too long', however, is open for contestation. Recently in New Zealand Alison Barrett, a 42 year old obstetrician and gynecologist, who is clinical director of maternity services at Waikato Hospital, revealed in a large full page newspaper article that she still breastfeeds her daughter Matilda who is nearly five (Irvine 2005: 7). Barrett admits she has been 'in the closet' (see Chapter 2 on coming 'out of the closet' in relation to pregnancy) about the fact she has fed her daughter for so long. Barrett has four children ranging from a 15 year old son to Matilda who is nearly five years old.

Barrett 'went public' about breastfeeding her daughter in order to put pressure on Parliament's health select committee to create a law on breastfeeding that sends a message breastfeeding is acceptable and should be encouraged. Currently, there is no law against breastfeeding in public or at work in New Zealand but anecdotal evidence suggests that 'many mothers say they are treated as if there is' (Irving 2005: 7).

> The Human Rights Commission has a long history of handling discrimination complaints and inquiries about breastfeeding. In a document called The Right to Breastfeed, published in February this year, the Commission says that from mid-2000 to mid-2004, it had 29 inquiries on matters relating to breastfeeding. Such inquiries are apparently increasing, and most involved women being prohibited from breastfeeding in public places, including high schools, restaurants or at community events. Some also involved an employer resistant to the idea of a woman expressing milk or taking breastfeeding breaks at work. (Irving 2005: 7)

Giles (2003) in a chapter entitled 'Weaning Frankie, age five and a half' explains that many 'standard' [read Western middle-class] mothering texts recommend weaning before the age of one, although the World Health Organization recommends that children be breast-fed until they are at least two. Although children in many cultures are breast-fed until they are much older (possibly as old as seven) 'for most Westerners, an older child at the breast in considered an offensive sight' (Giles 2003: 207–208). Giles received written responses to her questionnaire that indicated that women who feed 'older' children often retreat to the privacy of their homes to feed. One respondent wrote 'As the babies got to one and two and three years old, it was socially more and more unacceptable and I received lots of adverse comments like "That's disgusting"' (Giles 2003: 208).

Cindy Stearns (1999: 319) notes that the motivation for breastfeeding an older child is often questioned because it is thought that breastfeeding beyond infancy is not nutritionally necessary.

> Was the mother still breastfeeding because she could not 'let go' and did this thwart her child's need for independence? Was a woman breastfeeding for her own sexual and/or emotional needs as opposed to the child's? Breastfeeding women were aware that breastfeeding an older baby or child in public was often met with increased suspicion by strangers, family, and friends. Furthermore, children who are older can often talk and this raises problems controlling the disclosure of breastfeeding. In contrast to a baby who signals hunger with a cry that might mean bottle or breast, a child who can ask for its mother's breast alerts those present that there is a breastfeeding relationship. (Stearns 1999: 319)

Women face mounting pressure to discontinue breastfeeding as a child grows older. Stearns (1999) argues that one way women resist this pressure is to develop code words to conceal extended breastfeeding. She provides an example of where the code word 'milky milk', or another case 'nummies', is used to describe breastfeeding. This is understandable given the prevailing cultural norms about the appropriate duration of breastfeeding. Stearns (1999) provides an example from one of her research participants, Mary, a 36 year old mother of a nursing three year old:

> I had another friend who breast-fed probably 'til two or two and a half and her daughter got very, she just did demand all the time, you know, not any kind of regulation and so every time anything went badly, wrong, you know, distressing, Lily would start yelling 'titty, titty' and it'd be like in the middle of the restaurant or you know all over the place. So that's why we call it 'milky milk'. (Stearns 1999: 319)

As Stearns (1999) points out these code words do not change people's attitudes towards nursing children but they do allow mothers and children to maintain their breastfeeding relationship despite cultural pressures.

Medical experts continue to argue that 'breast is best', however, in many Western countries breastfeeding rates remain low. The disjuncture between breast for sexual pleasure and breast for nurturing infants leaves many women feeling uncomfortable about breastfeeding, especially in public places. Grosz (1994) argues women's bodies are constructed as a 'mode of seepage'. I think that this possibly plays a crucial role, at the level of the conscious and unconscious, in decisions to bottle rather than breastfeed, especially in public space which has long been associated with Rational man whose body has historically been coded as solid and in control.

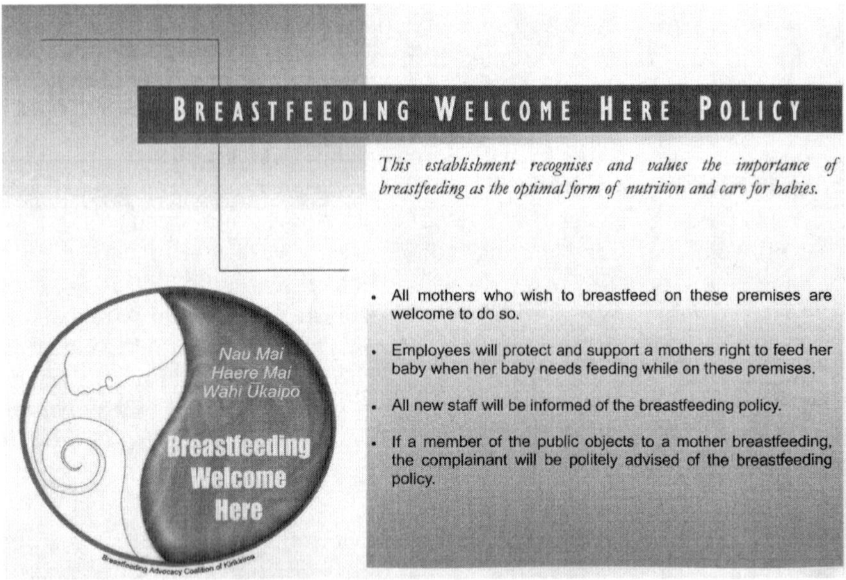

Figure 7.4 Breastfeeding Welcome Here Policy. *Source*: Breastfeeding Advocacy Coalition of Kirikiriroa (Hamilton).

The constriction of spatial freedom that breastfeeding women (but not men) experience has been documented by geographers (e.g. Mahon-Daly 2002; Pain *et al.* 2001) and others (Stearns 1999). Some people feel uncomfortable when women breastfeed in public space. This is perhaps not surprising given it is gendered performance that can cause 'gender trouble' (Butler 1990). At some times, in some places, for some bodies, breastfeeding is culturally sanctioned. At other times, in other places, for other bodies it is culturally condemned. The cultural and social practices of breastfeeding, therefore, need to be understood as spatially constituted. Recently in Hamilton, New Zealand, the Breastfeeding Advocacy Coalition distributed posters to restaurants and cafés stating 'Breastfeeding Welcome Here' (see Figure 7.4). The aim is to encourage women to feel comfortable breastfeeding in public space.

Rebecca Johnson (2005) draws attention to how lactating mothers make visible some of the persistent gender and sex-based exclusions from social spaces (see also Johnson 2004). Johnson begins with an account of her exclusion from an English pub based on her accompanying eight week old breast-fed baby. The child was a 'minor' and therefore not allowed on the premises as it would be breaching the pub's liquor license. Johnson (2005) argues that the breasted body is a boundary site that can be used to examine the role of regulatory legal texts in constructing gendered legal spaces. Johnson troubles the supposedly neutral category of 'minor', noting that there is no distinction made in legal texts between the minor who is eight weeks old and 18 years old and that with its focus on the 'minor', the legal text obfuscates and denies the interpenetrations of the bodies of children and adults (Johnson 2005: 5).[5] Breastfeeding is one particular relationship in which boundaries are blurred, and interpenetrations are part of this relationship. Johnson (2005) argues that through the legislation that denies the minor entrance to the pub, legislation also functions to deny entry to those who have continual relations with children.

'A SWIRL OF SQUEAMISHNESS'

Giles (2003: 133–134) comments 'breast milk seems to have been unnecessarily caught up in a swirl of squeamishness ... The fact that it comes out of a woman's bosom, and not a cow's udder, shouldn't reduce its appeal, from a rational point of view, and might increase it when considering its compatibility to the needs of the human body.' Discourses about bodily fluids, contamination, and abjection construct the ways in which breast milk is understood in many Western contexts. Natalie Angier (1999: cited in Giles 2003: 133) in *Woman: An Intimate Geography*, poses the question 'If you saw a full glass of human milk in the refrigerator, would you drink it? The idea is disturbing. It feels almost cannibalistic.'

Giles relays stories of women, who on occasion, used breast milk because they were out of cows' milk. One woman wrote to Giles (2003: 134) 'My mom once put some expressed breast milk into her minister's tea when he called and she didn't have any [cows'] milk and was so embarrassed that in desperation she used her own. He said it was the nicest cup of tea he'd ever had!' Another woman commented to Giles that she had once, when making a pie, had brushed the crust with breast milk and then sprinkled it with sugar. She was out of cows' milk and so simply expressed some over the crust. The woman claimed she had used breast milk in other recipes too but nobody ever seemed to notice. The idea of adults drinking or cooking with breast milk is, for many, strange and nauseating (but for those who are interested Giles includes at the end of her book a section on recipes that contain breast milk — they include Annuska's Sourdough Bread, Gayle's Breast Pump-kin Pie and Nick's Vanilla Diazepam Ice Cream).

There is more at stake here than the risk of infection through transmission of a body fluid. Giles (2003: 133) argues 'the risk of being infected by breast milk is negligible.' Breasts, especially lactating breasts, speak to a politics of fluidity. Breastfeeding is seen to poignantly mark women's 'difference' from, and Otherness to, men (although as the examples of male lactation indicate this embodied difference is not as marked as it might appear).

Sarah Walker, a social anthropologist in the United Kingdom posted a notice on the Bornfree website explaining that she is currently undertaking research into the sexual politics of breastfeeding. She says that she first became aware of men's capacity to breastfeed when she read Helen

Marieskind's (1973) article entitled 'Abnormal Lactation' in the *Journal of Tropical Pediatrics and Environmental Child Health*. Marieskind puts forward a scientific case for men's capacity to breastfeed. This case, says Walker

> undermines one of the key premises underlying much feminist literature on breastfeeding i.e. breastfeeding is the exclusive capacity of women. The fact that man's capacity to breastfeed is left largely unrealised in our society has far-reaching theoretical and practical implications and may be seen to reveal a great deal about our culture. (Bornfree, 2005, available <www.unassistedchildbirth.com>, accessed 3 June 2005)

I agree with Walker. What are the implications of aligning men with fluids and fluidity? Luce Irigaray (1985a) argues that in Western philosophical models of ontology, fluids, fluidity, and the viscous are subordinated to that which is concrete and solid. Fluids, fluidity, and the viscous are implicitly associated with femininity, maternity, pregnancy, lactation, menstruation and the body. Solidity is associated with masculinity and rationality. 'Solid mechanics and rationality have maintained a relationship of very long standing, one against which fluids have never stopped arguing' (Irigaray 1985a: 113). Irigaray is not necessarily proposing that women's bodies are inherently fluid while men's bodies are solid, but that men's and women's bodies become coded in this way. Male lactation recodes men's bodies as fluid, nurturing, and maternal. Not only women's breasts, but some men's breasts become reconfigured as sites/sights of (potential) flow of colostrum (the earliest form of milk that women can produce at approximately 28 weeks pregnant) and milk.

CONCLUSION

Breastfeeding is not simply a natural phenomenon. It is cultural performance. At some times, in some places, for some bodies, breastfeeding is sanctioned. At other times, in other places, for other bodies, it is condemned. The cultural and social practices of breastfeeding, therefore, need to be understood as spatially constituted. The bodies themselves are spaces, but these bodies/spaces are in turn located in other spaces. Breastfeeding can maintain the status quo but it also can rupture the status quo and change the social order depending on the cultural context. Breastfeeding as a sacred maternal performance is undone when men breastfeed, when breastfeeding is incorporated into people's sex lives, when lactating women feed animals, and/or when children (rather than infants) are breast-fed. These acts of lactation rupture the social fabric. Maternal breastfeeding is revealed as a performance that can be fractured.

Giles (2004: 302) argues that one of the most remarkable features of the material she collected when researching her book *Fresh Milk* was that the stories about lactation and the various ways in which it was taking place quietly in suburban kitchens and bedrooms, where mothers with surplus milk, for example, are adding it to the breakfast cereal, or feeding it to kittens; or mothers who are sexually aroused by breastfeeding are nursing their partners during sex. Yet the women and men engaged in these acts are ethically at sea. Beyond knowing intuitively that secrecy is wise, they have no idea whether they are alone in their perversity, or if everyone else is doing the same thing, and feeling quite normal about it, but just not telling.

There is a denial, concealment and erasure of these lactation activities. Giles (2004: 302) explains that she 'had unwittingly engaged in a process of queering, by opening out the closeted in a way that was both inquiring and celebratory.' Coming out of the closet either by choice or force can have repercussions. After the story of Kura Tumanako breastfeeding her puppy was published in the media, Tumanako was put under psychiatric care in the Hawke's Bay mental health unit. It is not known who is taking care of her three-month-old baby (Hodge 2004b).

My aim in this chapter was to 'queer breastfeeding' by opening up for discussion breastfeeding behaviors that sit outside the normative constraints that apply in contemporary Western culture. Lactation has long been a symbol of maternity — sustenance, warmth, goodness — and yet there is a vast and diverse range of lived experiences of lactation, some of which reiterate dominant discourses of the 'purity of mothers' milk' and others that contest these discourses. The examples of adult nursing, interspecies breastfeeding, men who breastfeed, and women who breastfeed children beyond infancy, illustrate that the social and spatial boundaries that surround breastfeeding shift. They show that the spatiality of lactation — where, when and whose breasts — is important and that in some instances lactating breasts are deemed to be in place but in many other instances they are deemed to be out of place.

8 'Bad' Mothers

(Re)presentations of Lack

Examining maternities and mothering opens up questions about what constitutes 'good' and 'bad'[1] (and the myriad of possibilities in-between) practice for mothers. Susan Chase and Mary Rogers (2001: 30) write:

> We all know the ideal of the good mother. Above all, she is selfless. Her children come before herself and any other need or person or commitment, no matter what. She loves her children unconditionally yet she is careful not to smother them with her love and her own needs. She follows the advice of doctors and other experts and she educates herself about child development ... The good mother's success is reflected in her children's behaviour — they are well mannered and respectful of others; at the same time they have a strong sense of independence and self-esteem. They grow up to be productive citizens.

Few mothers feel able to live up to these expectations. From the moment women conceive they are bombarded with advice on how to be good mothers. Every decision and action mothers take carries a moral weight. Chase and Rogers (2001) point out that although good mothers are depicted in terms of what they *do*, there are also unspoken assumptions about who they *are*. 'It is rarely said out loud that the good mother is a white, able-bodied, middle-, upper-middle-, or upper-class, married heterosexual' (Chase and Rogers 2001: 31). Chase and Rogers (2001) pose the question, if women who fit these criteria feel pressure from the good mother ideal then how much more pressure do those women who do not fit them feel? Good mothers and bad mothers also have varying access to different spaces and places.

What constitutes good and bad mothers also opens up questions about moral politics and panics as was seen in the discussion of Nikki 'porn mum' in Chapter 4 and of disabled mothers in Chapter 5. Some women are constructed as good mothers (usually white, affluent, able-bodied women who have a 'house and a husband') while others are constructed as bad mothers (usually non-white, poor women who have neither a house nor a husband) and this is often linked with moral judgments.

This chapter explores the experiences of some women who tend to be (re)presented as bad mothers (see Ladd-Taylor and Umansky 1998 on '"Bad" mothers: The politics of blame in twentieth-century America'). Mothers involved in the sex industry and disabled mothers have already been mentioned but others include teenagers, lesbians, drug users, addicts, alcoholics, violent mothers, ill mothers, single mothers, and mothers who raise children communally. Women who are considered to be poor, ill, uneducated, neglectful, and immoral do not enjoy a heightened status when they take on mothering roles, rather they tend to be denigrated at every turn.

This chapter is about those mothers who are deemed to be bad — to not live up to the ideals expressed above. There is no doubt that some mothers do behave badly. They abandon, mistreat, fail to protect and even kill their children. But this chapter is not about those mothers. It is about those mothers who are deemed bad on account of who they are and the social and spatial positions they occupy. It is not possible within the scope of the chapter to focus on all those mothers often deemed to be bad and so just three examples have been chosen: first, 'lone or single mothers' who receive Welfare or, in the case of New Zealand, a 'domestic purposes benefit' because they are unable to financially support themselves; second, teenage mothers who are thought to not be old enough to raise a child or children; and third, lesbian mothers who do not live in nuclear families. I do not seek to justify these 'other mothers' or to 'prove' that they can be good mothers. I take this as read. Rather the aim is to understand how and why discourses of bad mothering within specific social and contexts are constructed.

The chapter begins by theorizing lack. It is argued that bad mothers tend to be (re)presented as lacking in a number of ways. The chapter then discusses in turn mothers on Welfare, teen mums, and lesbian mothers. Finally, I examine a new 'geography of resistance' that has begun to emerge over the past decade — both the strategic occupation of, and resistance to, the categorization of 'bad mother'.

THEORIZING LACK

Bad mothers are often presumed to be lacking in some way. Lesbian mothers are thought to be lacking a man. Mothers on Welfare are thought to be lacking financial resources. Drug dependent mothers are thought to be lacking willpower and self-control to quit their drug habit. Mothers who do not live in nuclear families are thought to be lacking 'proper' families. Teenage mothers are thought to be lacking in maturity and mothering skills. It could be argued that the concept of 'lack' provides a useful lens through which to understand (re)presentations of all mothers, but I think it is especially useful for understanding (re)presentations of bad mothers.

In order to understand the concept of 'lack' in more depth it is useful to turn attention to psychoanalysis. For the purpose of this chapter, Lacanian psychoanalysis proved more useful than Freudian psychoanalysis. Steve Pile (1996: 92) explains:

> In Lacanian psychoanalysis, subjectivity is installed as a cover for the fear of loss, initially of the mother. The mother's body becomes a significant site/sight for the child. When it is seen that the mother lacks a penis, this confirms the reality of castration.

Boys and girls respond differently to anxiety about castration. Boys see that they have a penis and fear they might lose it. Girls see that they have no penis and assume they have already been castrated. Girls must come to terms with their loss (and their lack). In Lacanian theory this castration complex is understood not in biological terms but in relation to language. It is this psychoanalytic concept of lack in relation to *language* that is most valuable for understanding the construction of bad mothers as lacking. The realization of sexual difference, and of Woman as lack, is valuable for prompting a more in-depth understanding of gendered and sexed subjectivity.

> The most useful re-formulation of the castration complex for feminist discourse comes from Lacan, who shifts the focus from biology to differences produced for both sexes by language. Rather than connecting castration with the absence or presence of an organ, he uses the term to refer to alienation in language, to subjection to the symbolic codes of culture. (Wright 1992: 43)

Margaret Whitford (1991a and 1991b) makes the point that feminist theorists such as Luce Irigaray have explored the deep-rooted concept of 'Woman as lack' in Western culture. Irigaray

> begins with an analysis of the imaginary of Western philosophical and psychoanalytic discourse ... aiming to show that the conceptualization of sexual difference in this discourse is governed by an imaginary which is anal, that is to say, which interprets sexual difference as though there were only one sex, and that sex were male (women are defective men). (Whitford 1991a: 69)

This concept of Woman as lack pervades Western philosophy, logic, culture, symbols and social institutions. Lack is related to not just female sexuality but also to the ways in which Woman is positioned discursively. Amanda Banks (2005: 23) writes:

> This understanding of the concept of Woman has implications for access to language and participation within the cultural (read: public) sphere, and also shapes social institutions in such a way that Woman continues to be reinscribed as lack.

According to Lacanian psychoanalysis all mothers are coded as lacking but it seems that some mothers are coded as more lacking than others.

Banks (2005) conducted research on women who claim the Domestic Purposes Benefit (DPB) in Hamilton, New Zealand.[2] It was in part Banks's research that prompted me to think about the various ways in which some mothers are (re)inscribed as lacking (Burns 2000 also discusses how powerful ideologies of family and class construct Welfare-dependent single mothers as lacking). The DPB is a government benefit that is paid to single parents to cover the costs of raising children — the American equivalent of Temporary Assistance to Needy Families (TANF) (see Kingfisher 1996; 1999 and 2002 on women on welfare in the United States). Utilising feminist (re)interpretations of the psychoanalytic notion of lack (including Whitford 1991a, 1991b and Irigaray 1985a, 1985b) Banks (2005) argues that 'DPB mums' are seen to be lacking both materially and discursively: 'DPB mums' are seen to lack a 'proper' work ethic, a husband for herself and a father for her children. They are seen to lack, therefore, a 'proper' (nuclear) family.

Whilst recognising that the psychoanalytic notion of lack in some psychoanalytic texts relates mainly to anatomical lack — the girl's lack of a penis — Banks (2005) posits that the notion of lack extends beyond this to cover the ways in which Woman is positioned within social, cultural and discursive realms. Moira Gatens (1996: xii) also makes this point:

> whether or not it is coherent to speak of a woman's body as *anatomically* 'lacking', is irrelevant to the imaginary apprehension of women's bodies as 'begging the question' of their completion by a man and/or child (italics in original).

Gatens (1996) speaks to the connection between Woman as lack and heteronormativity. Woman, it seems, can only potentially be made whole or complete by Man and/or a Man and a Child. This helps explain the privilege that tends to be afforded to the heterosexual union which brings about a Child (which makes Woman 'complete'). Thus, in psychoanalytic terms, Man is constructed as 'complete', the reference point for female sexuality, whereas Woman is constructed as fundamentally lacking.

In articulating Woman's discursive construction as lacking, it is beneficial to understand how lack is created within dualistic systems of thought. A great deal of work informed by feminist and poststructuralist theory seeks to destabilise the dualisms that underlie Western philosophy (see Grosz 1994; Rose G 1993). Much of this work relies on insights from Jacques Derrida (1974, 1978) who developed the theory and method of deconstruction. Derrida is concerned mainly with the instability of language. His main premise is that

> our beliefs in reason are built through a series of unequally valued binaries ... the first term assumes priority and the second, inferior, term 'marks a fall' and is conceived in relation to the former as a complication, a negative, a manifestation, or a disruption of the first. (Johnston *et al.* 2000: 625)

Derrida argues that binaries can be understood not as *A* and *B*, but as *A* and *not-A* (Rose G 1993). Grosz (1994: 3) echoes this sentiment, stating that 'dichotomous thinking necessarily hierarchizes and ranks the two polarized terms so that one becomes the privileged term and the other its suppressed, subordinated, negative counterpart.' Not only is the first term valued over the second, but the debt to the second term is not acknowledged. The primacy of the first term relies upon it silencing the second term. Numerous feminists have extended this analysis by arguing that dualistic thinking is gendered. McDowell and Sharp (1999: 11) explain:

> women, and feminine characteristics, are defined as inferior, irrational, emotional, dependent and private, closer to nature than culture in comparison with men and masculine attributes that are portrayed as superior, rational, scientific, independent, public and cultured. Women ... are at the mercy of their bodies and their emotions, whereas men represent the transcendence of these baser features: mind to women's body.

As a way of disrupting dualistic thinking, Derrida advocates that a new term be developed, thus rupturing the binary.

The significance of this in relation to this chapter is that (re)constructing some women who have previously been considered to be bad mothers — as lacking rather than having — might only be a partial solution. It would not deconstruct the binary but simply reverse it, perhaps in the interim before a third term is found. Eradicating the thinking that constructs some mothers as good and some as bad, some as having and some as lacking, is what is required. This means rethinking politics of ethnicity, age, sexuality, disability, class and so on because the terrain of the bad mother is intersected and informed by these different axes of embodied subjectivity. For example, non-white women are seen to possess a rapacious sexuality through their supposed closer relationship with nature (see McClintock 1995). Ethnicity makes a difference to how the bad single mother who claims Welfare is constructed. The notion of 'the family' is also implicitly racialized and authors such as Heidi Nast (2000) have shown the conflation of the (nuclear) family with notions of whiteness.

The care provided by health-care professionals for bad mothers in pregnancy and after birth is hugely variable. Single mothers birthing often report being treated with disrespect, or provided with inappropriate care.

This chapter does not present a seamless story of mothers who lack but rather is a piecing together of textual fragments, of vignettes of 'different' mothers, mainly mothers on Welfare, teenage mothers and lesbian mothers who create cultural anxiety about their ability to nurture and rear a child or children. At the end of the chapter I return to make some more general comments about taken-for-granted assumptions about motherhood and the spaces mothers inhabit, alluded to in the introduction to this book.

LONE MOTHERS ON WELFARE

The composition of families in the West has undergone significant changes over the past 40 years: the incidence of divorce, single parenthood and 'blended' families has increased dramatically. Damaris Rose (1993: 194) says that mother-led families are no longer unusual. Even 20 years ago in Canada the census data (1986) showed that about one in six families with dependent children were headed by a female lone parent (Rose 1993: 194). Despite these changes over the decades, romantic notions of the 'ideal' family continue to be reiterated in the media and many social customs, with the 'traditional' (nuclear) family maintaining a discursive stronghold (Banks 2005).

Central to the notion of 'the family' is the figure of the 'good' mother. The influence of feminism not only altered the meaning of work for women (and indeed for many men) but it also signaled a concerted shift for women away from the private space of the home into the public workplace. One would think that perhaps with these widespread demographic changes there would be more discursive and material space for diversity in family life. However, particularly with the rise in divorce rates and the occurrence of single motherhood in many places, it is very difficult, if not impossible, for most women to measure up to the image of the good mother. Most women, rather than being perceived as having particular maternal attributes, are perceived as lacking. This is especially the case for mothers who receive Welfare.

Banks (2005: 26) lists a number of dualisms that construct mothers on Welfare:

Man – Woman
Nuclear Family – Single mother and children/'fatherless' family
Father – Mother
Have – Lack
Presence – Absence
Complete – Incomplete

Using both focus groups and individual interviews Banks talked with a total of 16 Pākehā (European) mothers. All lived in Hamilton, New Zealand, and all were on a DPB. Banks also interviewed two people who work at the local Work and Income New Zealand (WINZ) office. Banks positioned herself as an 'insider' in the research. She concluded:

> The notion of Woman as lack operates at material, discursive and psychical levels for Pākehā women claiming the DPB. These 'DPB mums' are interpellated as lacking and feel that the lack — particularly the lack of a husband-father and therefore 'proper' family — is a key aspect of how they are perceived as 'DBP mums'.

Banks' participants were especially aware of being perceived as lacking in 'family spaces' such as 'family restaurants'. The lone mothers in this environment felt acutely aware of not having a man/father with them to 'complete' their family. Some were also aware that there is a pervasive discourse in New Zealand that represents DPB mothers as fiscally wasteful and lacking in their ability to manage money effectively. 'People' think they should not enjoy 'luxuries' such as eating chocolate biscuits at home or eating out. But Banks' participants did not wholeheartedly accept this reading of their lives and resisted the dominant discourse that DPB mothers are lacking. Instead they talked about what they *had*. Many of the 'DPB mums' articulated diverse definitions of what constitutes a 'family' and positioned themselves as *having* a high degree of personal autonomy, thus challenging the notion that they were *lacking* through being a single parent claiming Welfare in New Zealand. Clearly power is not just about domination but also about resistance. As Joanne Sharp, Paul Routledge, Chris Philo and Ronan Paddison (2000: 1) point out, there are a 'myriad of *entanglements* that are integral to the workings of power' (italics in original).

'GIRL MOMS'

Teenage mothers are another group who tend to be perceived as lacking. This is evidenced on the website 'Girl.Mom' which offers support, community and education for young mothers. Many young mothers post their stories on the site. For example, Jenni Macri tells:

> When I sat in my bathroom looking at the pregnancy test, all my worst fears were realized. Two pink lines stared back at me. I was numb. The next few weeks were filled with conflicting emotions about the pregnancy. The things I knew about unwed young mothers weren't good things. I knew how I looked upon girls I knew that got pregnant and it wasn't how I wanted to be looked at. (Girl.Mom, 2006, available <http://www.girl-mom.com/index.php>, accessed 18 April 2006)

Macri says that the things she knew about 'young mothers weren't good' and that she herself was guilty of judging 'girls' who got pregnant. Teenage mothers in the West are commonly constructed as lacking. To make matters worse their age is often coupled with the fact that young mothers are also more likely to be poor, uneducated, part of an 'underclass' and have a lower health status than their middle-class, married counterparts (Bonell 2004).

It is important to keep in mind though that the notion of what constitutes a 'young mother' is culturally specific. In non-Western countries, young women from the age of approximately 12 years are sometimes married and involved in childbearing and rearing. Because of the separation of extended families in the West, and the ways in which the nuclear family is privileged as a child-rearing unit, many young women (and indeed men) often have scant experience in caring for children. Child rearing, in the West, is predominantly seen as an activity that is engaged in by parents rather than older siblings, as happens in many other cultures. But even within the West there are cultural differences. In New Zealand, for example, Māori and Pacific Islanders have higher birth rates than other ethnic groups. 'Nationally, Maori are almost five times more likely to have teenage births than Europeans, while Pacific Islanders are three times as likely' (*Waikato Times* 14 October 2006: D1). Despite these cultural differences, by and large, many families, teachers, doctors, midwives, nurses, and social researchers regard teenage pregnancy as problematical.

Chris Bonell (2004) embarked on a project to find out *why* teenage pregnancy and mothering is regarded by researchers in the United States and United Kingdom as a major social problem. Bonell conducted a review of published materials in the United States and United Kingdom between 1981 and 2000. The findings are intriguing. Research from the United Kingdom often identified teenage pregnancy as problematic because of health issues. Research from the United States often identified teenage pregnancy as problematic because of associated Welfare expenditure. 'A few studies from the USA regarded teenage parenting as mediating the intergenerational transmission of poverty. Such studies often focused on Black and minority ethnic populations' (Bonell 2004: 255).

Gail Reekie (1997: 77) examines the 'bodies of illegitimately pregnant women'. She explains that the body of the illegitimately pregnant teenager causes cultural anxiety or is 'particularly troublesome' because it is replete with contradictions. It is youthful but physiologically mature, childlike yet maternal, innocent but sexual, and the object of both compassion and censure (ibid). Reekie (1997: 77) continues: 'The body of the pregnant teenager exposes the limits of what is culturally thinkable about proper sex and motherhood.'

David Sibley (1995) points to the liminal spaces occupied by adolescents. He explains that they threaten adults because they transgress the boundaries between adulthood and childhood and 'appear discrepant in "adult" spaces' (Sibley 1995: 34–35). The 'girl-mom' likely poses a heightened threat especially to the 'adult-mom' in spaces such as antenatal classes, crèches, and play groups.

To return to the psychoanalytic notion of lack, the pregnant teenager is seen to lack one or more of the following: maturity, wisdom, financial resources, self-control, a husband, and an 'instinct' to mother since she herself is still considered to be a child. A recent newspaper article on 'young' mothers in New Zealand was entitled 'Kids having kids' (*Waikato Times*, 14 October 2006: D1). It tends to be assumed by many (not only lay people but also social researchers) that all teenage pregnancies are the result of an 'accident' and are 'unwanted'. Bonell (2004: 259) notes that no studies of teenage pregnancy from the United States and United Kingdom took into account the mother's assessment of whether her pregnancy or birth was wanted or intended. Teenage pregnancies from the outset are seen to be the result of a lack of self-control and self-determination.

LESBIAN MOTHERS

Another group of women who tend to be seen (especially by social conservatives and/or homophobes) as mothers who lack are lesbian mothers. Janet Kenney and Donna Tash (1993: 119) note 'In recent years, an increasing number of lesbian women have chosen to bear children.' Even so, lesbian mothers are still often regarded as 'lacking' — lacking a male partner (even though some may have a relationship with a man or with the child's father), lacking a 'proper' family, and lacking 'proper' morals. Resisting this notion of lack, some children of lesbian (and gay, bisexual and transgender) parents claim a positive identity as 'queerspawn':

> 'Queerspawn' is a term used by some sons and daughters of LGBT [lesbian, gay, bisexual and transgender] parents who claim and celebrate our identity, acknowledging that, regardless of our own sexual orientation, our heritage has 'queered' us. While there are millions of kids in LGBT-parent households — up to 14 million by some estimates — I'm guessing only a fraction of them would actually call themselves queerspawn. I'm one of them. (Abigail Garner, Queered by Family, Queerspawn by Choice, available <http://damnstraight.oversampled.net/2001/11/02/queerspawn-by-choice/>, accessed 18 April 2006)

Lesbian mothers raise children in a variety of family circumstances. Some care for children on their own or with extended families, some share parenting with male friends, husbands, or ex-husbands, some co-parent with a lesbian partner (Chase and Rogers 2001). Chase and Rogers (2001: 31) remark 'Estimates of the number of lesbian mothers in the United States range from two to four million. Nevertheless, in American society lesbian motherhood has been considered an oxymoron.' Being a lesbian mother is sometimes considered incongruous, therefore, lesbians can find that motherhood makes their sexual orientation invisible to others. 'Passing', at times, may be useful. What is not useful, however, is the assumption that lesbians are bad mothers or mothers who lack because they are not heterosexual. Victoria Clarke (2000: 273) comments that much of the feminist and psychological literature on lesbian mother families is concerned with the question 'are lesbian families the same as heterosexual families?' (and thus, 'are lesbians "fit parents"?'). Clarke reviews this often contradictory literature concluding that that there are things that are both the same and different in heterosexual and lesbian households.

Lesbians tend to be constructed as bad mothers because, like gay men, they are associated in the popular imaginary, with sexual activity. There is a long history dating back to the Victorian cult of true womanhood and domesticity that mothers are asexual and 'pure'. There is convincing evidence that good parenting is not related to sexual orientation (see Chase and Rogers 2001: 33–35) and yet cultural assumptions continue to construct lesbians as bad mothers. These assumptions underpin discriminatory laws and policies. Lesbians are often threatened with or denied custody of their children. Limits are put on visitation rights. Despite the long standing assumption that children and mothers belong together, lesbians often fear that when a marriage ends they will lose legal rights to their children. The lesbian mother's child-rearing skills or relationship with the child or children is not taken into account, rather it is presumed that the lesbian mother is a bad mother who is perverse and lacks sound moral judgment (Kenney and Tash 1993).

This positioning of lesbians as perverse and lacking, however, has been challenged. Jaqui Gabb (2004: 403) explains:

> while 'lesbian' is positioned as a deviant category, and in the cultural imaginary butch lesbians are identified as the corruptors of young girls, female sexuality is traditionally seen as 'safe', especially sex-less, self-less motherhood. Indeed, it has been suggested that child sexual abuse is a 'man thing' (Itzen 2000).

In this context, it has been argued that lesbian mothers offer the safest home spaces in which to raise children (see Saffron 1998 on some of the advantages of having a lesbian mother).

Gabb (2004) undertook research in Yorkshire, UK, between 1999–2001 on lesbian parent families. Gabb, herself a lesbian mother, interviewed 13 lesbian parent families in depth — 18 parents and 13 of their children. In an article titled 'I could eat my baby to bits: passion and desire in lesbian mother-children love' she writes:

> Lesbian mothers often sought to re-establish a sense of physical connectedness, through intimate (embodied) practices between mothers and children. In fact 'other mothers' ... appeared to use intimacy and the

> erosion of bodily boundaries as a means to develop their sense of unity and connectedness with their (non-biological) children. This requires careful 'family sexuality' management as the physical and affective boundaries between their roles become blurred. (Gabb 2004: 409)

Gabb is not implying that lesbian mothers' acknowledgement of their passion for their children is potentially sexual or incestuous but that 'mother love' can be intense.

Gabb's research prompted me to wonder given that dominant discourses construct lesbian mothers as lacking and as lesser versions of heterosexual mothers maybe lesbian mothers feel a need or desire to fill the perceived gap or lack with an intense physicality with their children. There is certainly room for more research to be done in this area.

Dominant ideologies of motherhood, reproduction and the family frequently represent women as bad mothers, or women who lack in particular ways. There are numerous examples. Just three — lone mothers on Welfare, teenage mothers and lesbian mothers — have been mentioned in this chapter. Mothering is a powerful construct that privileges some (such as 'white', heterosexual, able-bodied, middle-class and married women) and oppresses others. But those women who are constructed as Other by dominant Western discourses of mothering and families are not without voice.

'BAD MOTHERS CLUB': GEOGRAPHIES OF RESISTANCE

Over the past decade some women have taken the signifier 'bad mother' and claimed it as part of a positive identity in order to highlight the harsh judgments that women as mothers are continually subjected to. Such tactics of resistance are reasonably common. Turning labels such as 'queer' and 'nigger' into positive signifiers are examples that spring to mind. A Google Search in April 2006 for 'bad mothers' resulted in 27,200,000 hits. This number of hits is perhaps not surprising given the dominant discourses about bad mothering but what is interesting is that of ten listed sites on the first result page only three were links to 'serious' sites that discussed bad mothers. These included an article on 'Bad mothers' in the *Journal of Social History*, a website that claims 'Dr. Phil helps bad mothers' and a website on a university study of whether educated women make bad mothers.

The other link were to 'humorous' sites such as the 'Bad Mothers Club' (a web magazine for mothers who 'need a laugh, some solidarity with fellow strugglers or a reason to sneak away for a precious half-hour'); 'The F-Word — Bad Mothers' (a website constructed by Claire Riley who says she will never go 'warm, mushy and gaga' over motherhood, and that feminism needs to bust the 'perfect mother myth'; *The Bad Mothers Handbook* (Long 2005); and *Confessions of a Bad Mother* (Calman no date).

The website 'Bad Mothers Club' (2006, available <http://www.badmothersclub.co.uk/jsp/index.jsp>, accessed 12 September 2006) advertises a 'Bad Mother Doll' that shouts, swears, and buries her head in her hands. She also comes with laddered tights and lipstick smeared face. Consumers can purchase a brunette, afro or blond with black roots. Her head tilts back to hold a cigarette and vodka bottle. Bad Mother Room Sets are also available including the pub, club, Department of Social Security and betting shop.

Over the past couple of years there have also been films and television series that engage (often playfully and humorously) with the idea of bad mothers. One example is the BBC comedy series 'Absolutely Fabulous' (colloquially abbreviated to 'AbFab') which first screened in 1992. In 'AbFab' Jennifer Saunders plays the role of a jet-setting, bad-living PR agent and single mother Edina (Eddy). Julia Sawalha plays her strait-laced and disapproving daughter Saffron (Saffy). Edina is undoubtedly a bad mother who is far more interested in living it up with her chain-smoking, hard-drinking, drug-imbibing, school-days chum Pattsy Stone, played by Joanna Lumley, than she is in caring for her daughter.

Although in the West motherhood is glorified there is a virulent denigration of individual mothers and particular groups of mothers such as mothers on Welfare, teenage mothers and lesbian mothers who do not fit the ideal, who are seen to be lacking in some way. The websites and television series mentioned above can be read in multiple ways. They can be read as further putting down those women who are seen to not measure up, subjecting them to further derision. Some people may not find them at all amusing. However, the sites can also be read as women attempting to resist the dominant discourses of mothering. One site claims 'the Bad Mothers Club takes a much-needed swipe at the guilt-inducing perfect parent philosophies' (Families South East. Join the Bad Mothers Club, 2006, available <http://www.familiesonline.co.uk/article/articleview/443/1/45>, accessed 7 September 2006). Mothers have been blamed for numerous problems (see Ladd-Taylor and Umansky 1998). Television comedy series (e.g. AbFab), websites (e.g. Families South East. Join the Bad Mothers Club) and popularist books (e.g. Long 2005) are one way of striking back.

CONCLUSION

There are many taken-for-granted assumptions about what constitutes a good mother — a mother who has what it takes — which renders many other mothers such as lone mothers on Welfare, teenage mothers and lesbian mothers as lacking. Mothers on Welfare are situated as deficient through their lack of material resources, teenage mothers through their youthfulness and lesbian mothers through their lack of male partner. What constitutes 'good' mothering practices is narrowly defined. Those mothers who fit the criteria enjoy privileged access to a variety of usually middle-class or affluent spaces. They are thought to represent stability, order, and attachment to the mainstream values of mothering, that is, being loving, nurturing, self-sacrificing, morally pure and so on. Unlike bad mothers, these mothers do not provoke moral panic. They are seen to be the bedrock of proper 'families', societies, and nations.

In the next chapter I change tack. Rather than focusing specifically on various constructions of maternal bodies I examine instead a particular space — cyberspace. Chapter 9 addresses the rise of information on conception, pregnancy, birthing, and infant feeding in cyberspace. It heeds Rob Kitchin's (1998: 403) appeal to examine further the complex and mobile spatialities of cyberspace and builds on the important work of Madge and O'Connor (2005) on 'e-mums'. It also follows in the footsteps of Hester Parr (2002) who examines the rise of health and medical information on the Internet. Like Parr, I examine how this information on the Internet is communicated and used by particular groups. Electronic information flows have the potential to prompt social and cultural change. For example, chat rooms for women who are attempting to conceive, or women who are experiencing difficulties breastfeeding, can create a sense of belonging to 'communities without propinquity' (Webber 1963: 23 cited in Valentine 2001: 118). I am interested in the ways in which the Internet affects women's embodied experiences of maternity and vice versa.

9 ClubMom.com

Constructing Maternal Identities in Cyberspace

ClubMom is an online United States organization for 'moms'. The Internet site 'clubmom.com' informs readers that the club started 'because moms raise our nation's kids, strengthen our communities, drive our economy ... and you deserve to be recognized and rewarded for this' (ClubMom, 2004, available <http://www.clubmom.com>, accessed 11 July 2004). The site, which offers a 'resource centre', a 'community', and online shopping with 'rewards and savings' is typical of literally millions of Internet sites available to mothers-to-be and mothers, especially in the United States and United Kingdom. Thus far I have examined the connections between maternal bodies and a variety of material and discursive spaces but not cyberspace.[1] This is despite the fact that in writing this book I often found myself searching for information on the Internet.[2]

The chapter is based on research conducted online. The methodology involved gathering data from Internet sites and chat rooms. There are a plethora of sites and routes to finding information about conception, pregnancy, birth, infant feeding, and parenting on the Internet. I was helped immensely in thinking through how to approach this chapter by an edited collection of essays on *Doing Internet research: critical issues and methods for examining the net* (S. Jones 1999).[3]

Work by Madge and O'Connor (2005) also proved useful in researching and writing this chapter. They draw on empirical evidence from a study on the role of the Internet for new mothers to argue that cyberspace and geographical space coexist and that 'although cyberspace can result in the production of new selves, these selves have residual attachments to embodied experiences and practices' (Madge and O'Connor 2005: 83). Madge and O'Connor (2005) argue cybermothers live online and onsite lives simultaneously. They researched one specific Internet site — Babyworld <http://www.babyworld.co.uk> — and conducted a web-based questionnaire survey and semi-structured virtual group interviews with site users. Madge and O'Connor (2005) call the Internet space, 'cyber/space'. They explain 'The use of the backslash in cyber/space is intended to be emblematic of the mutually constituted, dyadic nature of both spaces, where clear distinctions between the two spaces can become blurred' (Madge and O'Connor 2005: 83).

Before proceeding, however, it needs to be acknowledged that not everyone has computer access. Parr (2002: 78) explains 'virtual geographies are still relatively exclusive, with only some individuals having the ability to access these spaces.' Many women in the world manage to conceive, carry their babies in utero, give birth and nurture babies and children without ever turning on a computer. This is an important point that arguably deserves more attention but in this chapter I am unable to address all the issues surrounding differences in access to Information and Communications Technology (ICT). For now, the focus is on maternal bodies that *do* interact with virtual space and the socio-spatial effects of these interactions, but the issue is raised because it is important not to lose sight of the fact that many mothers-to-be and mothers do not have ICT access.

The chapter is structured in five sections. First, a review situates the Internet search on maternities within the geographical literature on virtual space. Second, a brief account is given of the online methodology used to elicit data for the chapter. Third is a discussion of one of the key themes that emerged from the research — cyberspace as an arena for online shopping. Following this is a discussion of another key theme — information and advice on maternity available online. The final theme that is discussed is the Internet as a site of social action. Women often exchange and circulate information through bulletin boards and chat rooms with a view to improving the rights of and services for mothers.

BODIES AND CYBERSPACE

This section is a review of some of the interdisciplinary work that has been carried out on the burgeoning topic of cyberspace. Others have already reviewed this work (e.g. Kitchen (1998a) and more recently Madge and O'Connor (2005: 84–86)) so the comments here are brief and aimed more at providing a context for understanding constructions of the social and spatial relations surrounding maternities and mothers online rather than charting this huge territory more broadly.

Kitchen (1998a) argues spatiality is vital to understanding cyberspace. He introduces cyberspace to an audience of geographers and lays out the debates and empirical research being engaged by scholars in other disciplines. Since the publication of this work, many geographers have begun to explore constructions of social and spatial relations in cyberspace.

Gill Valentine and Sarah Holloway have published a large body of research on 'cyberkids' (e.g. Holloway and Valentine 2001a, 2001b, 2001c; Valentine and Holloway 2001a, 2001b, 2001c, 2002). Much of their research draws on primary empirical data on children aged 11–16 and their parents' experiences of, and attitudes towards, ICT. Valentine and Holloway (2002) examine the incorporation of children's 'real' and 'virtual' worlds. They argue there are four main ways in which this occurs. First, children use online activities to reconfigure offline relationships. Second, children use ICT to find information about their offline hobbies and interests. Third, children make 'virtual' friends who may then become offline friends. Fourth, ICT can recontextualise children's offline identities in negative and positive ways.

Hester Parr (2002) examines online geographies of health arguing there has been a shift in emphasis from the state provision of medical care of the body corporate in spaces such as doctors' surgeries, clinics, and hospitals, to the individual body who 'consumes' medical information in spaces such as holistic health shops, private health clinics, gyms and telephone help lines (also see Nettleton and Bunton 1995). Parr (2002: 73) uses a Foucauldian understanding of medicalisation, 'the notion of a "fourth spatialisation" of the medical gaze' in order to examine the Internet as a 'new geography of health promotion which enables human subjects to "self diagnose" and to "discipline" their bodies.' She collected empirical examples from multiple sclerosis (MS) chat rooms where people came together in virtual space to discuss medical knowledges which surround 'ill' bodies (also see Crooks 2006 on chronically ill women's use of online health information).

Research on the formation of online communities in cyberspace has begun to proliferate over the past few years. Different authors have explored different axes of identity. In an interesting article entitled 'India.com' Paul Adams and Rina Ghose (2003: 414) focus attention on ethnicity and the ways in which temporary and permanent immigrants to the United States from India use virtual space or what Adams and Ghose refer to as 'bridgespace' to 'supports flows of people, goods, capital and ideas between South Asia and North America.' The concept of 'bridgespace', while intended to describe flows across a geographical divide, helped me to think through some of the issues surrounding the bridging that occurs when one moves from a life without maternal responsibilities to a life with maternal responsibilities.

This section does not come close to indicating the significant contributions made by geographers and others over the past decade to examining the diffusion and adoption of information technologies, cyberspace, virtual reality, telematics, the information society and so on (e.g. Adams 1997, 1998; Crang *et al.* 1999; Cutchin 2002; Hillis 1996; Janelle and Hodge 2000; Shields 1996; Taylor 1997). Instead, I have simply outlined just a few of the published works on cyberspace and communities that I found most useful for the purposes of thinking through issues surrounding maternal bodies and cyberspace.

VIRTUAL METHODS

Laura Gurak and Lisa Ebeltoft-Kraske (1999: 147), in their introduction to a special issue of *The Information Society* on gender and computer-mediated communication (CMC), argue for research on CMC to move away from 'attempts to generalize and toward studies that are situated in specific contexts and specific technologies. It is no longer possible to comment on *all* CMC, because there are too many contextual factors involved' (italics in original). I think there is merit in Gurak and Ebeltoft-Kraske's claim but in relation to becoming and new mothers still relatively little seems to be known about their use of the Internet — how they gather, disseminate, and exchange information on mothering. Therefore, the task of formulating at least some kind of general picture of what cyberspace looks like — its flows, contours, and configurations — for becoming and new mothers is important. This chapter, therefore, limits discussions to a particular 'discourse community' and context (Gurak and Ebeltoft-Kraske 1999: 147), that is those who have an interest in mothering, but it has not been limited to a particular online communication. I have examined Internet sites, online chats, online shopping, and bulletin boards to ascertain what information is available to those interested in finding out more, or connecting with others, on topics such as conception, pregnancy, birth, breast- and bottle-feeding, and parenting.

With the aid of a research assistant, Amanda Banks, I conducted a series of searches (mainly using Google but also some other search engines including xtramsn) for key words including conception, fertility, pregnancy, baby, breastfeeding, bottle-feeding, mothering, toddler, child care, and parenting. These searches generated millions of hits. For example, a Google search for 'conception' generated 15,400,000 hits, 'pregnancy' 26,200,000 hits, 'birth' 67,000,000 hits, and 'breastfeeding' 3,260,000 hits. Amanda and I spent many long hours, over a period of more than a month, trawling through literally hundreds, if not thousands, of Internet sites, checking bulletin boards, looking at 'baby galleries', reading birth stories, messages, newsletters, advice from experts, and answering quizzes. This 'survey' makes no pretence towards sampling based in the quantitative tradition. Many of the Internet sites we visited offered no way of discerning how many users were logging on, etc. Rather, we tended to visit sites in a random fashion clicking on those we thought looked 'interesting' for a variety of reasons such as they linked to other material covered in this book, they were well-known and/or large sites, they were unusual and so on. We did not systematically select every, say, fifth site on a listing.

Amanda and I had a number of discussions about our findings and the impressions we were left with from our random visiting of Internet sites over the period of several months. We decided that the sites we visited could be divided into three broad groups (which sometimes overlapped): first, sites that provide online shopping; second, sites that provide information and support on issues to do with maternity; and third, sites that prompt some kind of social action. My reflections on these explorations of cyberspace, informed by discussions with Amanda, are presented in the remainder of the chapter.

SHOPPING ONLINE

Madge and O'Connor (2005) claim that virtual space and material space are concurrently entwined and mutually constituted. Perhaps one of the most powerful ways in which this occurs is that both are embedded in global capitalism (see Shields 2003 Chapter 7 on virtual business). Virtual space as a technology creates discourses which in turn fashion maternal bodies in particular ways, in this instance, as consumers. Although consumerism surrounding maternal bodies is not new (mothers have long purchased products such as a crib, pram, clothing, nappies, and bedding in order to 'prepare' a room for a new baby) in recent times a greater array of goods and services, both onsite and online, have become available. The Internet appears to be saturated not only with information for mothers-to-be and mothers but also with goods and services available for purchase.[4]

Some Internet sites have changed over time introducing retailing after recognizing a marketing opportunity. For example, the well-known site 'Babyworld' began by offering advice and an online community for parents but in February 2000 they launched one of the UK's biggest online retail shopping sites (Babyworld, 2005, available <www.babyworld.co.uk/shop>, accessed 10 July 2005) for pregnant women and parents of infants. Many of the Internet sites that Amanda and I visited are 'sponsored' and have links to product lines available for purchase. As I have noted elsewhere (Chapter 3 this volume and Longhurst 2005a) in the 1990s the advertising and retailing industries began marketing products to pregnant women and new mothers. 'Advertising and retailing industries ... are always looking for new markets and while in the 1980s marketers began selling the idea of beauty, fitness, and health to young men (Miller *et al.* 1998; Mort 1996), in the 1990s marketers began seeing pregnant women as offering commercial possibilities' (Longhurst 2005a: 437). Consider, for example, the 'Robin Rive' range of teddy bears which are sold online (Bears from Robin Rive, 2006, available <http://www.bearshops.co.uk/manuf04.html>, accessed 12 September 2006). The range includes 'Preggy Teddy' who is attired in a green smock with a floral pattern, trimmed in lace. The bear has a large bulging stomach that contains a baby bear (the pregnant stomach or pouch is zipped between the mother bear's legs). The market for Preggy Teddy is likely to be collectors of bears, but also pregnant women, new mothers, and young children. I was given a Robin Rive 'Preggy Teddy' when I first began research on pregnancy in the early 1990s.

Another example of online retailing is the vast array of maternity wear now available. Some of this maternity wear is typical of that worn in the 1950s and 1960s — smock tops with ample shirring or gathering — or in the 1970s to 1990s — 'tent dresses' and baggy T-shirts — but most is fashionable and aims to show 'the bump' rather than hide it. As discussed in Chapter 3, years ago pregnant women were expected to borrow maternity wear (see Gregson and Beale 2004 on the circulation of maternity wear) and to shop instead for the coming baby. This appears to be changing as pregnant women over the past decade have become increasingly fashionable. A Google Search for 'maternity wear' returned over five million hits. there are sites such as 'Maternity 4 Yummy Mummies,' 'EziBuy Maternity Wear,' and 'Mummy Maternity' that advertises itself as offering a large range of maternity clothing and accessories such as casual wear, office wear, formal wear, swimwear, pregnancy piercings, and maternity wedding gowns (see Mummy Maternity, 2006, available <http://www.mummymaternity.co.nz/>, accessed 12 September 2006). The Mother Store (2006, available <http://www.themotherstore.co.nz/>, accessed 12 September 2006), which is New Zealand owned and operated, offers a wider range of products to women who are planning to conceive, pregnant or have just had their baby. They sell online skincare and aromatherapy including midwife created pregnancy remedies, designer nappy bags, 'Preggie Pop Drops' (sweets to ease morning sickness), a screensaver called 'ExpectantSee' that grows with your pregnancy, 'de-alcoholised wine', 'Peanut Shell Slings' worn by celebrities including Angelina Jolie (these slings cradle the baby across the adult's body), a liquid holster that clips on to the side of prams or strollers, a portable highchair and many more goods.

But it is not only pregnant women and new mothers who offer commercial possibilities. So too do women and their partners who want to become parents. For example, Internet sites on conception offer for purchase natural remedies, prenatal vitamins, synthetic hormones such as Clomid that stimulate follicle growth to enhance fertility, ovulation predictors and pregnancy test kits. Male and female (in)fertility has become big business since the 'baby boomers' started delaying childbearing and found it difficult to conceive.

There are two possible reasons why so much online shopping is available for mothers-to-be and new mothers. One is that pregnant women, women with babies, and breastfeeding women often find it difficult to negotiate public spaces. Maternal bodies are frequently constructed as forgetful, overly emotional, and irrational — modern day hysterics (Longhurst 1997).[5] In turning to the index of nearly any book on pregnancy it does not take long to come across a heading such as 'emotions' or 'emotional changes'. Schrotenber and Solomon Weiss (1985: 125) write: 'The emotional life of a pregnant woman has been compared to a volcano: All that's been hidden beneath the surface, lying dormant under the tranquility and the everyday concerns of our adult lives, bursts forth in an eruption of feelings and fears.'

But not only are pregnant women frequently constructed as forgetful, overly emotional, and irrational but also as dangerous, leaky, and unattractive bodies that break their corporeal boundaries (see Grosz 1994 on women's bodies as 'modes of seepage'). These discourses make it difficult for women to occupy public space. It is widely assumed that people have to possess consummate control over their bodily/mental functions in order to walk confidently in public space (Longhurst 2001b). Both the material and symbolic realm function to exclude pregnant women (although it seems that in Hamilton, New Zealand, this is now less so than a decade ago — see Longhurst and Banks 2004).

It can also be difficult for parents of infants, toddlers and young children to occupy public space. As Valentine (2001) points out children are often presumed to be less civilized than adults and less able or willing to follow rules of conduct. It can be exhausting shopping with a baby, toddler or young child. Constant attempts to discipline young bodies so they behave 'appropriately' in public spaces such as high streets and shopping malls can be a challenge for all involved — care-givers and children. Purchasing goods and services online rather than onsite may prove easier for some mothers.[6] Madge and O'Connor (2005: 85–86) make the point 'It is possible that ... cyberspace may be particularly significant for new mothers because women's embodied experiences are commonly denied expression in the symbolic order.'

Some mothers may make some purchases in 'material' space (or 'geographic space' to use Kitchen's 1998 term) and other purchases in cyberspace. Kitchen (1998: 403 cited in Madge and O'Connor 2005: 89) explains: 'cyberspaces do not replace geographic spaces, nor do they destroy space and time. Rather cyberspaces coexist with geographic spaces providing a new layer of virtual sites.' In relation to new mothers Madge and O'Connor (2005: 89) note 'On the whole, online interactions supplemented and added dimensions to the women's lives rather than copying or replacing the "real world" scenario.'

A second possible reason why there is such a proliferation of online services and products available may be that consumption is a way of forging a new identity as a mother. Becoming a mother involves a range of performative acts, one of which is purchasing items (possibly for the first time) such as disposable nappies or diapers (Huggies, 2006, available <http://www.huggiesclub.com/>, accessed 2 October 2006), parenting magazines (Mother and baby, 2006, available <http://www.motherandbabymagazine.com/> accessed 2 October 2006), toiletries for mothers and infants (Boots, 2006, available <http://www.boots.com/>, accessed 2 October 2006) and strollers and infant car-seats (Strollers.com, 2006, available <http://www.strollers.com/>, accessed 2 October 2006). Madge and O'Connor (2005:

83) argue new mothers explore cyberspace 'as a performative liminal space' where they can 'try out' different versions of motherhood. They discuss this trying out of different maternal identities in relation to women being part of virtual parenting communities — visiting favorite Internet sites and exchanging information online with other parents — but these different versions of motherhood may also be expressed through processes of global online consumption — buying maternity wear for a special event such as a wedding, selecting a feeding bra, or choosing a crib. These acts are not solely about the materiality of the goods being purchased but they also position women discursively in the role of mother.

INFORMATION, ADVICE AND ONLINE 'COMMUNITIES'

Not only is there now a vast array of goods available online but there is also a vast array of information about health, medications and how to manage illness available online (see Crooks 2006). The Internet also offers chat rooms, support groups, bulletin boards, and a huge range of information on subjects such as conception, pregnancy, birth, lactation, infant feeding and parenting. Given that mothering is such a profound, consuming and life-changing event for many it is perhaps not surprising that it provokes the exchange of a vast array of information. Elsewhere I have argued that pregnancy is an important 'rite of passage' (see Teather 1999) and one facet of this rite of passage is pregnant women giving, and, more usually, receiving advice (Longhurst 1999).

> In particular, women who are pregnant for the first time seem to receive an enormous amount of advice as they make the transition to motherhood. This advice comes from a range of people including health workers, friends, acquaintances, loved ones, colleagues, and even strangers. It also comes from a range of sources including pamphlets, manuals, books, newspapers, magazines and advertisements. The advice often focuses on topics such as diet, exercise, birth, medical procedures, lactation, how to care for a newborn baby and how to raise a child. (Longhurst 1999: 78)

When I look back at this excerpt I am surprised I did not include the Internet in the line-up of sources of information and advice for pregnant women. Nearly ten years ago when I wrote this, online information on pregnancy and mothering was readily available but not to the extent it is today. Over the past few years there has been something of a revolution in the uptake and use of information on the Internet. However, it is not just pregnant women who seek, receive and offer advice online but also women attempting to conceive, and mothers of newborns, toddlers and older children. The Internet is a source of information and advice for literally millions of people on various aspects of mothering.

Madge and O'Connor (2005) base their research on 'e-mums' on one of the UK pioneer parenting websites 'Babyworld', launched in 1995 (available <http://www.babyworld.co.uk>, accessed 10 July 2005). They note that in May 2001 the site had a membership of approximately 50,000 (p. 93). When I accessed the site in July 2005 the membership was approximately 60,000. The site claims to have 1.5 million hits a month and over 2,600 pages of advice are available on a range of issues.

It is important to note in relation to the sites such as 'Babyworld' that they offer a source of information or advice that mothers and others can actively seek rather than passively 'receive'. The advice is sought rather than unwanted. Subjects exercise their agency when they log on to the Internet in order to seek or exchange information. In this regard the Internet can be seen as fulfilling a positive function. Sometimes pregnant women and new mothers can find the experience of receiving advice to be rather overwhelming. They can experience a sense of being under surveillance, and of being regarded not as a subject in their own right but as existing solely to serve the interests of their fetus, baby, toddler, or child. The 'giving' of advice can be read as an attempt to impose limits on mothers' bodies and behaviors. Sometimes mothers tire of receiving advice from friends, family, colleagues, and even strangers in the street but if they log on to the Internet to search for advice it can be assumed, to some degree at least, that this advice is being sought and is therefore welcome. This is not to deny that some women may feel societal pressure to log on particular sites because they want to be seen as being 'knowing subjects' in relation to mothering.

The Internet, however, is a space that does more than simply provide information. The information also constructs maternal bodies in particular ways. Many of the Internet sites examined for the purposes of researching this chapter provided advice and guidance that reiterated hegemonic discourses of health and medical care. Birth, for example, is constructed on many sites primarily as a medical/hospital based event. There are sites that offer information on water-birth, acupuncture and 'hypnobirthing' but these tend to be represented as 'alternative' birthing methods. Normative ideas about birth are strongly represented not just in material space but also in virtual space. Most online health advice sits within a Western scientific medical framework, for example women are advised to increase their intake of folic acid and avoid coffee during pregnancy. Madge and O'Connor (2006: 199) also note that 'very traditional stereotypes of mothering and gender roles' persist on the Internet. There is a lot of information about nutrition, and about charting the changes in a pregnant woman's body and how the baby/fetus grows and changes throughout each week of pregnancy. Facts and figures about conception, pregnancy, birthing and lactation inscribe women's bodies through medical discourses (see iVillage.

co.uk, 2006, available <http://www.ivillage.co.uk/pregnancyandbaby>, accessed 2 October 2006 as an example of a UK-based 'website for women' that has special pages on 'pregnancy and baby' with message boards that illustrate the topics commonly discussed by those interested in maternities in the 'cyber community').

There are some Internet sites, however, that contest Western, mainstream, scientific, masculinist paradigms of medicine, health and social development. A useful example is provided by Internet sites on midwifery. Maria Fannin (2006) argues that individual midwives and midwifery advocates are taking advantage of Internet chat rooms, websites and discussion lists to create linkages that stretch across national boundaries. They are creating a kind of 'global political midwifery movement' (p. 70). Fannin (2006: 172) examines these 'invocations of "global" midwifery emerging out of activist networks in North America' concluding they have the potential to establish new forms of connection but these forms of connection are complex.

SOCIAL ACTION: IMPROVING 'A MOTHER'S LOT'

Fannin's (2006) research on 'global midwifery' provides a useful lead into the third theme that emerged as Amanda and I trawled through hundreds of Internet sites on maternities, that is, social action. In short we navigated our way through a variety sites that aimed to improve 'a mother's lot'. Ruth Panelli (2004: 184) says: 'Social action can be understood as the acts, practices and strategies people implement individually or collectively to maintain, modify or challenge the structure/s and/or operation/s of places, settings and societies in which they live.' There are many sites that aim at improving the rights of and services for mothers, fathers, children and families in general.

Perhaps one of the best examples of the way in which cyberspace is used for social action in relation to maternities is the NCT site (National Childbirth Trust, 2006, available <http://www.nct.org.uk/>, accessed 2 October 2006). The NCT is a UK-based charitable trust that has been in existence for approximately 50 years. They claim to have over 60,000 members across the United Kingdom. The Trust functions in material space and cyberspace. Their website explains:

> We are the leading charity dealing with pregnancy, birth and early parenthood in the UK, operating at both local and national levels across Scotland, England, Wales and Northern Ireland ... We work with governments and policy makers to ensure that parents' needs are represented and included in policies and guidelines at both national and local levels. (National Childbirth Trust, 2006, available <http://www.nct.org.uk/>, accessed 2 October 2006)

The Trust is also involved in education and offers courses for parents, both before and after birth. They also research and produce evidence-based information which they claim enables parents to make informed choices. In addition to this they deliver practical training sessions for the National Health Service (NHS) to 'enhance and develop the skills of health professionals, midwives, health visitors and all those involved in maternity, neonatal and postnatal services' (National Childbirth Trust, 2006, available <http://www.nct.org.uk/> accessed 2 October 2006). They have a network of NCT trained breastfeeding counselors, antenatal teachers, postnatal leaders and NCT representatives working with their local maternity units and in partnership with other local organizations. The Trust claims that this enables them to provide services that address the needs of parents in diverse communities as well as promoting and enabling higher standards of local maternity services to benefit all parents. NCT argue that they also campaign to protect the rights and choices of all parents in the United Kingdom and influence the attitudes of society. Local branches across the United Kingdom are run by volunteers. The success of NCT in providing support for the rights for *all* parents is no doubt debatable but it does seem NCTs presence on the Internet has provided space for at least imagining and reconfiguring relationships with greater numbers of people and exchanging information across wider symbolic and material divides.

There are also groups such as the US-based, not-for-profit organization, Childbirth Connection, formerly Maternity Center Association.[7] They claim to have worked to improve maternity care for mothers, babies and families since 1918 and to promote safe, effective, and satisfying 'evidence-based maternity care' (Childbirth Connection, 2006, available <http://www.childbirthconnection.org/>, accessed 11 October 2006). Like the NCT they too have a strong presence on the Internet. The Childbirth Connection's philosophy is that in the United States pregnancy and birth are commonly treated as medical conditions rather than as normal life processes and yet most women are likely to be well and healthy and can expect to have an uncomplicated birth and a healthy newborn. They argue for 'evidence-based maternity care' which means 'using the best research about the safety and effectiveness of specific tests, treatments, and other interventions to help guide maternity care decisions' (Childbirth Connection, 2006, available <http://www.childbirthconnection.org/>, accessed 11 October 2006). Social action for the Childbirth Connection involves research, education and advocacy with a view to improving the standard of maternity care.

These are just two examples of Internet sites that can be read as constituting a form of social action for mothers. These two charitable trusts — the NCT in the United Kingdom, and Childbirth Connection in the United States — engage in a range of online and offline activities. Both groups have been in existence (in one form or another) for more than 50 years and so represent examples of long-term social action that have been fostered in part by developments in information communication technologies. Neither group is looking to *radically* change the social fabric of society (see Fluri 2006 on marginalized groups using Internet technology to attempt to prompt revolutionary change) but they are looking to improve the social, political and economic conditions under which the work of mothering is performed.

There are also a range of other sites aimed at improving maternity care, some of which encourage shorter term social action such as commenting on a particular bill or act of parliament that affects mothers and/or families, signing a petition on a specific issue, supporting particular campaigns, writing or e-mailing to lend weight to a group action, subscribing to an e-newsletter and so on. Issues such as adoption; child support; child custody; unpaid domestic labour; mothers' rights in workplaces; and the wage gap between mothers, women who are not mothers, and men are common topics of discussion and action.

CONCLUSION

After months of searching and viewing hundreds of various Internet sites on maternities I find myself not wanting to either celebrate or condemn cyberspace on this topic. On the one hand it is possible to argue that to some degree at least ICT provides for some qualitatively new configurations of social and spatial power. Mothers can 'try out' different identities online. They can establish, through roles as consumers, by exchanging advice with others, or by becoming engaged in some form of social action to improve the rights of mothers, new ways of making meaning. ICT offers a space that brings together mothers from different spaces — rural spaces, cities, regions and countries. Cyberspace supports flows of maternal bodies, goods, information and social action across communities, towns, cities, regions and nations.

As Adams and Ghose (2003: 416) note, however, 'New technologies play a clear role in accentuating social divides and favoring pre-existing authority structures.' What becomes obvious, when trawling through hundreds of Internet sites and message boards, is that cyberspace is dominated by Western women. What is more representations are mainly of white, middle-class women and babies. The kind of cultural practices, information and advice (re)presented on conception, pregnancy, birth, and parenting are eurocentric and 'mainstream'. It is of course not surprising that Western women's narratives about motherhood have the highest profile in cyberspace given that these are the women with the highest access to computers and the most purchasing power.

Finally, ICT does not overcome or render insignificant material expressions of social, economic and political power, rather online and offline spaces are mutually constituted. The 'technosocialities' (Crang *et al.* 1999: 5) that exist amongst mothers in cyberspace are similar to those that exist in material space. The three major kinds of cyberspace that emerged when we (re)viewed hundreds of Internet sites reiterate spaces that exist in geographical space — mothers as an identifiable and important target for producers, retailers and marketers; mothers as receivers and providers of advice; and mothers as agents for social change. As Madge and O'Connor (2005: 83) suggest there is a 'simultaneity of online/onsite worlds in cybermothers' lives.'

10 Conclusion
The Contradictory Spaces of Mothering

The literature on maternities, maternal bodies and mothering is vast but there is still a gap in this literature — a gap this book has attempted to fill. Little research has been conducted on the relationship between subjectivity and spatiality. Issues of space, place, territories, borders and boundaries in relation to maternities, maternal bodies and mothering have not yet been explored in great depth. The purpose of this book, therefore, was to attempt to make a contribution in this area. I wanted to unravel some of the diversity and complexity of the embodied experiences of maternity and illustrate that maternal bodies are constructed through different social and cultural entanglements and through different spaces and places. The aim was to make a contribution to the geography literature by addressing previously ignored bodies and spaces. It was also to make a contribution to the maternities literature by offering a spatial perspective. In each of the chapters, therefore, I have filtered maternities and maternal bodies through a spatial lens. The book was not designed to offer a comprehensive account of maternities but rather a series of windows or views that prompt new ways of thinking about what it means to mother in particular spaces and places.

Throughout the book the focus has rested on maternal subjects rather than on children even though without one the other can make no sense. For too long, however, mothers have been marginalized in discussions of maternity. As Jolly (1998: 2) notes 'Although the mother might appear to be the central subject of maternity, she is often evacuated from this position by a discursive focus on the child.' In this book I have attempted to put maternal subjects centre stage in an attempt to understand more about the diverse embodied subject positions and spaces they construct and are constructed by.

This final chapter draws together some thoughts on coming out as pregnant at work, clothing the pregnant body, being physically disabled and pregnant, the politics and spaces of birthing, 'queer breastfeeding', representations and constructions of bad mothers, and mothers online in order to illustrate that maternal bodies and spaces are neither autonomous nor coherent. Instead I want to make an argument for recognizing the deeply contradictory practices surrounding the subjectivities and spaces of mothering.

Judith Butler (1990: 33) in her influential book *Gender Trouble* argues that gender is 'the repeated stylization of the body, a set of repeated acts within a highly rigid frame that congeal over time to produce the appearance of substance, of a natural sort of being.' Mothering is a specific kind of gendered performance that is also 'a set of repeated acts within a highly rigid frame.' It is time to 'trouble' that frame, that is, to 'do mothering' in a range of 'differently' embodied ways and to change the 'regulatory practices' that govern mothers' bodies.

Too often motherhood is seen within an essentialist and narrow frame. This book is an attempt to recognise different kinds of mothering that happen within a range of different spaces. Distinctions between mothers, fathers, and those who are considered 'childless' in relation to the care of children need to be reconsidered. It is important to decouple maternity and women's bodies not in an attempt to evacuate mothers as the central subjects in maternity but in an attempt to de-essentialize mothering and increasingly extend nurturing roles to men and to others.

It is possible for men to raise children, to nurture and care for them in ways that might be described as maternal or motherly but the question remains, can they 'really' be mothers? Thomas Laqueur (1990; 1992 cited in Aitken 2000: 582) points out that the 'biological facts' of motherhood and fatherhood are not 'given' but come into existence through political and cultural contestation. Neither discourses about maternities, nor 'real' maternal bodies, can be taken for granted. Consider, for example, male lactation (as discussed in Chapter 7). Consider also, Carlo the pregnant man. The following story, entitled 'Pregnant man thrilled to be giving birth', was reported in the New Zealand newspaper *Waikato Times*:

> 'Carlo' the pregnant man is about to give birth — probably next month.
> The 32-year-old male nurse, who his Philippines' doctors say is six months pregnant, is an hermaphrodite born with both male and female organs.
> The man, whose name has not been made public but who has been given the nickname Carlo, says he is thrilled and delighted about the prospect of motherhood. (*Waikato Times,* May 27, 1992, 1, cited in Longhurst 2001a: 5)

More recently, the website 'RYT Hospital — the first male pregnancy' (available <http://www.malepregnancy.com>, accessed 14 April 2006) invited visitors to 'join physicians and scientists around the world in monitoring Mr. Lee's pregnancy online.' It is possible to access Mr. Lee's 'vital statistics' such as his weight, blood pressure and fundal height and to see an ultrasound scan of the baby.

Lactating and pregnant men do not (yet) make sense to most people. Throughout the book I've stressed the point that maternal bodies only make sense within specific geographical and temporal contexts. The diverse range of online comments on Mr. Lee's pregnancy illustrate this point. Liz from the United Kingdom says: 'This is against God. Mr. Lee is a sick man. Repent now before it's too late. You sick, sick people.' Anne from the United States says: 'Go, Mr. Lee! All my prayers and best wishes to you. I'm so happy for you and your baby ...' Bob from Senegal comments: 'I've dreamed about this my entire life. Sign me up!' Clearly, opinion on the issue of male pregnancy is divided. 'New biotechnologies — such as cloning, sex selection, surrogacy, ectogenesis and gene manipulation — mean that pregnant bodies, and perhaps in the future this will increasingly be 'male' bodies, are a site of continual negotiation and contestation' (Longhurst 2001a: 5).

Bodies are undoubtedly a complex amalgamation of fleshy materiality and discourse. I'm not sure if the pregnant Mr. Lee is 'real' or an elaborate representation. In some ways, though, maybe it doesn't matter because regardless he makes possible a discursive connection between male embodiment and maternity. Clearly, in most cases it is *not* men who conceive, carry in utero, birth, and breastfeed babies. However, it would be amiss to simply ascribe these biological processes to women as though women are the only ones capable of being maternal bodies. Mothering is a multi-faceted configuration of socially constructed relationships that are played out on and through different material bodies.

Another example of the way in which mothering is a complex configuration of socially constructed relationships articulated through different bodies is the Couvade syndrome. The word 'couvade' comes from the nineteenth century French word *couver,* meaning 'to hatch' (*Collins English Dictionary* 1979: 345). It has come to mean a man having a 'sympathetic pregnancy'. Men tend to display some of the 'symptoms' (such as vomiting, weight gain, and changes in sexual appetite) experienced by their pregnant partners (see Klein 1991; MacClancy 1993). Some men today wear 'sympathy bellies' to simulate their partner's pregnancy.

Discourses of male maternal embodiment are becoming more common. For example, in the movie *Meet the Fockers* (the sequel to the 2000 comedy *Meet the Parents*) the character Jack Byrnes (played by Robert DeNiro) invents the 'Mannery Gland' that allows him to 'breastfeed' his one-year old grandson. In the movie 'TransAmerica' a pre-operative male-to-female transsexual embarks upon an unforeseen journey when she learns that she fathered a son, now a teenage runaway hustling on the streets of New York. The point here is that the work of mothering is complex in both its materiality and discursivity. It cannot easily be attached to any one essentially female body. Andrea Doucet (2006) in her book *Can Men Mother?* presents data collected from interviews with over 100 fathers in Canada to argue that some men are radically re-visioning child care and domestic responsibilities.

There are those who are likely to feel that if men share the discursive terrain, and sometimes even the fleshy materiality, of maternity it will rob women of the power bestowed upon them for their capacity to conceive, birth, feed and raise children. But while some take pleasure and pride in women's capacities to be 'goddesses of fertility', others take pleasure in being able to share the work of mothering with others, including men. Many (but not all) societies tend to lack respect for the nurturing and caring work often carried out by women in domestic spaces (Johnson with Huggins and Jacobs 2000). Men taking up more of this work may shift certain assumptions that women's lives and roles are somehow lesser than men's. Men mothering has the potential to shift hegemonic understandings of maternities. Children too are likely to benefit from having opportunities to receive and offer love to a variety of people who might take up maternal responsibilities in their lives.

This does not mean that I necessarily support the views and actions of 'Unions of Fathers' or 'Fathers' Unions'.[1] Fathers' Unions tend to argue for men's rights to *father* not to mother. They tend to argue that a child has a right to have both parents — a mother and a father — in their lives. Fathers' Unions, drawing on essentialist understanding of fathering, often promote 'the value of fatherhood' (Union of Fathers, 2006, available <http://www.uof.org.nz/>, accessed 8 December 2006). What I am arguing for, however, is that it is possible and indeed valuable for fathers (and others) to *mother*.

Such a move to 'de-essentialize' maternity is not aimed at nullifying women's 'difference' and the special value of pregnancy, birth and lactation but at recognizing the layering of meanings attached to maternity. How else is it possible to make sense of the fact that some men embody desires to mother and that some women do not? How is it possible to understand the hurt felt by a physically disabled pregnant woman when people express outrage about her decision to become a mother? Or how can one comprehend fully 'Nikki's' decision to be filmed giving birth for a pornographic movie? What is at stake in one adult breastfeeding another?

These questions illustrate that what it means to be a mother is not fundamentally tied to the role women play in physical reproduction. This is not to imply the biology — the flesh, blood and bone of women's reproductive bodies — is not important. Rather it acknowledges that women's bodies, mothers' bodies, are situated within particular discourses that surround motherhood. Maternity is not natural, innate, or essential. It cannot be taken for granted that women who have given birth will be 'good' mothers who love their children unconditionally and meet their children's needs selflessly day after day. Maternity is shaped, and reshaped, over time and in different spaces. Often, however, one version — a Euro-American version — is prioritized and others are expected to adopt this version. Jolly (1998: 1) explains:

> In many Asian and Pacific countries indigenous patterns of mothering have been challenged and to some degree transformed, first in the colonial period and second in the postcolonial epoch, in the name of civilization, modernity and scientific medicine. But women in these places

> have not simply succumbed to the message of 'enlightenment' through maternal improvement, but have variously rejected or embraced such advice or accommodated it selectively.

Jolly continues that this rejecting or embracing different modes of mothering is not like choosing goods from a supermarket shelf, rather women's choices are situated within complex gender, ethnicity and class relations. A complex politics inhabits maternities. Maternal bodies are spaces but they also inhabit spaces — psychoanalytic, discursive and material spaces.

The argument to 'de-essentialize' mothering opens up space for other researchers to address maternities in a variety of different contexts. For example, a great deal of work remains to be done on the roles played by kin such as grandparents, aunts/uncles and siblings when biological mothers are deemed 'unfit' (they may be incarcerated), are chronically ill, or have passed away (in recent years countless mothers in sub-Saharan Africa have died of HIV/AIDS leaving children who have become known as 'AIDS orphans'). I have focused heavily on some of the issues surrounding men mothering but the argument could also be taken in other directions.

A concept that is useful for thinking about 'de-essentializing' motherhood and the inseparability of bodies and spaces is 'corporeal space' (Moss and Dyck 1999). Moss and Dyck (1999: 389) refer to 'corporeal space' as a space 'where the discursive and material are synchronous.' The idea conveys something of the complexity, richness, but also the messiness, of space. 'Corporeal space' is understood to be contextual, discursive, material, economic, biological, physiological, fluid, and 'congealing from time to time around the body, only to be destabilized with new boundaries forming when any part of the context, the discourse, or the materiality shifts' (Moss and Dyck 1999: 389).

Patrice DiQuinzio (1999: 248) argues for a 'paradoxical politics of mothering' (also see Rose G 1993 on 'a politics of paradoxical space'), that is, for:

> A politics of mothering that accepts the impossibility of motherhood and the impossibility of individualist subjectivity, and that does not require for its foundation a univocal, coherent, and exhaustive position on mothering.

I agree with DiQuinzio. The wide variety of issues and spaces related to pregnancy, birth and child rearing covered in this book indicate that mothering is not coherent and consistent. There are dominant discourses that reflect and reinforce particular hegemonic kinds of mothering but these are continually subverted and rearticulated in a multiplicity of different ways.

A woman subverting the masculinism of her workplace by coming out as pregnant, a fashionable woman revealing her pregnant body in a public space, a woman being filmed giving birth for a pornographic film, and a new mother becoming part of an e-community of other new mothers are all examples of the complex politics inhabiting the multiple spaces that mothers occupy. People need to be able to mother — to offer care to children — in a variety of different ways. The distinctions between mothers, fathers, and people who are 'childless' are not at all rigid. All persons can (but not all want to or should have to) participate in mothering. Some mothers want to work full time in paid employment, others do not; some mothers want to breastfeed in public, others do not; some mothers want to appear sexy, others do not; some mothers are lesbian, others are not. Material and social support needs to be offered to *all* mothers who require it, not just to those who fit the image of the 'perfect mother' by conforming to dominant ideological forms of mothering.

Not only it is important, however, to re-imagine the politics of mothering but also the places and spaces of mothering. In this book I've focused on some of the spaces that tend to be typically associated with maternity such as homes and birthing wards, but I've also attempted to include other spaces less commonly associated with maternity such as brothels (in which breast milk is offered to clients), workplaces, welfare spaces, city centres, and cyberspaces. Examinations of pregnant, birthing, and child-rearing bodies should not be confined to the domestic realm because maternity is articulated in *all* places and spaces. Rewriting maternal bodies and spaces means examining women's and men's experiences of maternity and examining a variety of spaces.

I am not meaning to suggest that maternity is simply played out on a variety of different stages but rather that it is articulated *in* and *through* a range of bodies and spaces. One particular practice of mothering (such as breastfeeding a five year old child) might make sense and be accepted in one space (at home) but not in another (a café). Where a woman births, whether it be at home, in a birthing facility, or in a hospital, matters. Hopefully, the series of windows on maternities, gender, bodies and space offered in this book have revealed the multiple ways in which mothering and maternal bodies are imbued with social and cultural meanings. These meanings cannot be neatly pinned down. Continual attempts to secure mothering within particular boundaries of what constitutes normative femininity are bound to ultimately fail. Fortunately these boundaries cannot be secured. Discursive reiterations of what constitutes mothering change over time and space and are always open to contestation.

Appendix
Research Methods

Chapter 2 and Chapter 3 are based on the same data set. The data presented in these two chapters were derived from a study of first-time pregnant women who live in Hamilton, New Zealand. Nineteen first-time pregnant women were recruited to participate in the project. These participants responded to an article on the proposed research published in the local newspaper (*Waikato Times* 2002).

The women who contacted me were not screened in any way except that they were pregnant for the first time and that they were visibly pregnant. Preliminary interviews in 1992 prompted me to interview only women who were visibly pregnant because in all the interviews, others' reactions to the women's pregnant bodies, once people 'could tell', were discussed. Once again, I decided to stick with this decision in choosing participants for the 2002 study. Twenty one women contacted me and of these 16 were interviewed. Three more women were interviewed using snowballing from the initial contacts. Each woman was interviewed just once. The interviews lasted between 30 and 90 minutes. The interviews were semi-structured and described to participants as focused on women's engagement with dress, activities and places during pregnancy. Fifteen of the interviews were conducted in participants' homes, two in cafés, and two in my office at the university. The questions did not focus solely on workplace relations or maternity wear and body image but extended to various activities the interviewees had continued, reduced or stopped during pregnancy and places/spaces they occupied during pregnancy.

Each of the interviews was taped and transcribed in full. Hard copies of the transcripts were put in a folder. Detailed and repeated readings of them allowed me to identify recurring themes. I noted both similarities and differences in the topics that were discussed. For example, for the purposes of Chapter 2 I noted that reactions to the women's pregnancies at work varied greatly and so I grouped these reactions under different kinds of workplaces such as masculinised and feminised workplaces. For the purposes of Chapter 3 I noted that in nearly every interview some discussion emerged about receiving clothes from other mothers. I searched for networks, regularities and patterns in the data and developed 'hunches' about patterns that I thought were emerging, for example, that pregnant women were concerned about looking 'fat' in the first trimester. ('Trimester' is a common term used to describe different stages of pregnancy. The first trimester is from conception to the end of week 13, the second trimester is from week 14 to the end of week 26, and the third trimester is from week 27 to the end of the pregnancy.)

To aid interpretation of the interview data that follow, some information on the participants may be helpful. The 19 participants ranged in age from 17–43 (the average was 28 years). Sixteen identified as Pākehā (European), one as Māori, one as Māori and European, and one as a Pacific Islander. On average the participants were in their third trimester, or more specifically, 31 weeks pregnant (most babies are born between 38 and 42 weeks) at the time of interview. None identified as 'disabled'. All had experienced reasonably good health during their pregnancy. Two had separated from their baby's father during pregnancy, while the rest were in heterosexual partnerships. Sixteen were still engaged in full time employment but the women had strongly contrasting financial situations.

For the purposes of Chapter 3, in addition to the 200 pages of transcripts that were the main source of data, I also compiled a folder of maternity wear brochures, advertisements for maternity wear, newspaper clippings, magazine photographs and stories (mainly from *NZ Woman's Weekly*, *NZ Woman's Day* and *New Idea* and information from Internet sites selling maternity wear/fashion. In relation to the visual images of pregnant women I noted the similarities and oppositions between the images paying attention to whom the women were with and their pose, facial expressions, clothes, jewelry, hairstyles, and make-up. My analysis for both Chapter 2 and Chapter 3 was guided by Gillian Rose's (2001) discussion of discourse analysis of visual images because the relationship between 'visual images, verbal texts, institutions and social practices' (Rose 2001: 140) was my central concern.

Finally, in the course of my everyday interactions, over a period of several months while conducting interviews, I observed pregnant women in public and semi-public places such as at shopping malls, in workplaces, supermarkets, and swimming pools. Most participants were interviewed in their homes but I was also interested in how pregnant women perform their selves in public so wanted to observe what they were wearing, where they were shopping, who they were with, and what activities they were engaging in. At times, watching pregnant women stirred memories of my own pregnancies. In a sense, I relived the bodily sensations of when I was pregnant and reflected on what I had worn and how I had felt about my body and my clothing. In Chapter 3 I do not discuss these data explicitly but they provided a useful context for the research.

List of interviewees: conducted Hamilton, New Zealand 2002–2003

Name	*Weeks pregnant*	*Age*	*Occupation*
Mihi	28	17	Student
Nia	36	26	Natural therapist
Trish	41	28	Teacher
Lana	30	18	Former café worker
Eva	22	31	Sales assistant — dairy (groceries)
Jennifer	30	32	Scientist – public sector
Cathy	30	32	Registered nurse
Angelina	27	32	Radio announcer
Kate	36	30	Teacher – primary school
Sandra	35	26	Insurance claims assessor
Olivia	34	32	Sales assistant — nursery plants
Denise	33	29	Personal assistant
Sarah	27	25	*Whanau kaitiaki* (family support)
Hilary	22	28	Home loans consultant
Jo	30	28	Occupational therapist
Diana	35	32	Marketing manager
Carla	32	43	University lecturer
Dimi	31	29	Insurance claims clerk
Joy	34	27	Former office worker

Notes

CHAPTER 1

1. In addition to Third Space, Bhabha (1994: 38) uses a number of other spatial metaphors to articulate his notion of hybridity — these include 'alien territory', the 'split-space of enunciation' and 'in-between space'. I do not discuss these spatial metaphors here but nonetheless it is interesting to note that all of these terms could be applied in some way to the pregnant body.

CHAPTER 2

1. Much of the information on waged work and pregnancy pays attention to ensuring that pregnant women are not excluded from the workplace, that is, that they are afforded the same rights as other workers (even though these same rights might be dependent upon pregnant women's embodied difference). Lise Vogel (1990) reports on a ruling of the US Supreme Court in January 1987 in a case that 'posed the question of whether it is possible to reconcile equality norms with policies treating pregnant workers differently from other workers'. This case prompted divisions between feminist attorneys as to the 'merits and dangers of providing special benefits to pregnant workers' (Vogel 1990: 9). The arguments became polarised in terms of whether pregnant women should be treated equal to or different from other workers. Vogel (1990: 10) attempts to 'move beyond the polarisation that characterized the debate without losing sight of the larger political context'.
2. The origin of this saying is actually 'mmmmm is the word', the humming sound made with a closed mouth. Used by Shakespeare in Henry VI, Part 2. 'Seal up your lips and give no words but mum' (Phrase finder 2006 available <http://www.phrases.org.uk/meanings/211850.html> accessed 1 November 2006).
3. Preliminary interviews for an earlier study of pregnancy carried out in 1992-1994 indicated that first-time pregnant women tend to be more acutely aware of their changed corporeality as they begin to confront what it means to be a mother so I chose to select participants in the same way for the 2002 study. This enabled me to compare and contrast data from the two studies although I do not do this explicitly in this chapter.
4. Hamilton, located in the central North Island in New Zealand, is small by international standards having a population of just 166,128 (Statistics New Zealand, 2003). It is, however, New Zealand's fourth largest city. The CBD

is made up of shops (including several maternity outfitters), offices, cafés, bars, clubs, restaurants, galleries, a museum, and several cinemas, including a multiplex cinema. Design, style and fashion are influenced by trends in larger, more densely populated New Zealand cities such as Auckland and Wellington and international cities such as Sydney and London.

CHAPTER 3

1. These are three of New Zealand's top 'women's magazines'. *NZ Woman's Weekly* has been published for over 70 years. The magazine tends to focus on fashion and cooking. *NZ Woman's Day* has a similar focus. *New Idea* presents information on fashion, beauty, life, 'gossip', sex, health, relationships and Hollywood (see New Zealand top media-magazines (2005) available <http://www.business.vu.edu.au/bho2250/Top20Media/TopmediaNZ.htm#Magazines> for circulation and readership rates of these magazines).
2. For a list of women who have appeared pregnant and for the specific details of these appearances see Tyler (2001: 81).

CHAPTER 4

1. The subject of disabled women's sexuality has been largely ignored. Ann Neville-Jan (2004) reports that until recently there has been a gap on the topic. Disabled people's sexuality tends to be 'defined by deficit and sexuality either is not a problem, because it is not an issue, or is an issue, because it is seen as a problem' (Shakespeare *et al*, 1996: 3 cited in Neville-Jan 2004: 114). Neville-Jan (2004: 144-115) reports, however, that researchers in the US and the UK are now challenging these negative views of disabled people's sexuality.
2. It is somewhat ironic that while pregnancy appears to release some physically disabled women from being perceived as disabled, for able-bodied women the opposite often occurs. Pregnancy can make some able-bodied women feel as though they are disabled and in need of continual advice (Longhurst 1999). Pregnancy is sometimes represented as a disability rather than a 'normal' function in an able-bodied woman's reproductive life.
3. One of the implications of new reproductive technologies and procedures such as amniocentesis is that they have opened up a window on the womb which allows parents to ascertain if their baby might be disabled. Keith Sharp and Sarah Earle (2002) have interrogated the feminist and disability movements on the question of abortion. They conclude there are some fundamental and irreconcilable differences between these movements. Janice McLaughlin (2003) on the contrary, identifies some common ground in feminist and disability movements in relation to the social and ethical implications of antenatal screening and abortion.
4. For about a quarter of a century breastfeeding had gone almost completely out of vogue. It was a common saying that good cow's milk was better than inferior mother's milk. Truby King, however, argued that 'every young mammal is entitled to the milk specially designed for it by the Creator — the whale's milk for the baby whale, rabbit's milk for the baby rabbit, cow's milk for the calf, and mother's milk for the human baby' (Snowden and Deem 1951: 11).

5. Snowden and Deem (1951: 12) say:

> Their Excellencies, Lord and Lady Plunket, became keenly interested in the Society. Lady Plunket, in addition to lending her name to the Society, threw herself into lecturing and demonstrating, and thus extending the work to other parts of the country. So the 'Society for the Promotion of Health of Women and Children' now commonly called the 'Plunket Society' was formed.

6. In the Official Handbook of the Plunket Society (Deem and Fitzgibbon 1953: 63) mothers were advised to wake between 5 and 6 a.m. and to take a glass of hot water, a cup of weak tea, or orange or grapefruit juice. The baby was then to be changed, fed and put back in his [sic] cot to sleep. If there was no 'help' in the house then the father might prepare the morning drink, change baby and bring him to his mother. At 7.15 a.m. mother was to rise and at 8 a.m. she would have breakfast. At 9 a.m. babies would be given a sun bath if it was old enough and weather suitable. It was then to kick (as exercise) in its cot. At 9.30 baby was to be bathed and dressed and at 10.00 a.m. it was to be given fish liver oil, fed, and held out for bowel action. S/he was then to be put in cot to sleep in the open air or on a balcony or verandah. Detailed instructions such as these were provided for the entire day and night until 9.30 p.m. when the baby was to be given a hot milk drink, fed, changed and tucked down for the night.
7. When my first son was born in 1991 and I received a Plunket Book for him, my mother presented me with my own Plunket Book which she had kept in a drawer for 29 years. At the same time my partner's mother gave his Plunket Book to him.

CHAPTER 5

1. The phrase 'the West' is in quotation marks to problematise the cohesiveness or presumed singularity of 'the West'. 'The West' and 'the East' are not binary terms but often exist one within the other.
2. Hamilton is the site of Waikato Hospital which is a specialized 600 bed base hospital. The hospital has a staff of approximately 2500. It contains birthing facilities. Approximately 3000 women give birth at Waikato Hospital each year (Waikato Hospital 2005 available <http://www.waikatodhb.govt.nz/WDHB/default.asp?content=9> accessed 28 November 2005). Women in Hamilton can also choose to give birth at the Waterford Birth Centre or River Ridge East Birth Centre which are publicly funded, privately operated birth centres for low risk births, or at home.
3. Crow is director of Vixen Direct, New Zealand's largest pornography distribution company, based in Auckland, New Zealand.
4. Even without Crow's video there is still plenty of pornography featuring pregnant women (but not births). 'A quick survey of current print, video, and online pornography proves that pregnant chicks are actually a pretty popular fetish. Nearly every other issue of Hustler's Taboo … the hardcore fetish magazine from Larry Flynt's company, features a pictorial of a pregnant pin-up' (Taormino 2002: 1). A sexual fetish for pregnant women and birth is known as maiesiophilia.

5. Nikki and Crow both seemed to assume, at the early stages of the unfolding of events at least, that Nikki would give birth vaginally. Discussions about how birth by caesarian section might or might not 'work' in the film were absent in media reports.
6. See Binnie and Valentine (1999) for a review of 'geographies of sexualities' — also see Bell and Valentine (1995), Brown and Knopp (2003), Elder (1999) and Nast (1998).
7. Other newspaper articles that were not referenced in this chapter but informed the research and maybe of interest to readers are:
   Court allows porn film mother to keep baby (12 October 2002) Available <http://80tiki.knowledge-basket.co.nz> (accessed 6 November 2002).
   Mum's health upsets porn birth plans (23 October 2002) Available <http://onenews.nzoom.com/onenews> (accessed 16 February 2005).
   Porno birth mother visited by CYFS (31 August 2002) Available <http://onenews.nzoom.com/onenews> (accessed 16 February 2005).
   Porn birth woman asked to front up (19 July 2002) Available <http://onenews.nzoom.com/onenews> (accessed 16 February 2005).
   Porno birth filming up to doctor (13 October 2002) Available <http://onenews.nzoom.com/onenews> (accessed 16 February 2005).
   Porn film maker intends to shoot birth in private hospital (15 October 2002) Available <http://io.knowledge-basket.co.nz.ezproxy.waikato.ac.nz:2048/magz/007-01.html> (accessed 10 August 2005).
   Porn star gives birth prematurely (30 October 2002) Available <http://onenews.nzoom.com/onenews> (accessed 16 February 2005).

CHAPTER 6

1. Māori is the term commonly used to refer to the tangata whenua (literally 'people of the land') or indigenous peoples in Aotearoa New Zealand. I use this term here but wish to problematise such use. As Paul Spoonley (1993: xiii) points out, 'the word 'Māori' is really a convenience for Pākehā to lump together divergent groups.'
2. Hapūtanga is the Māori word for pregnancy. It is frequently shortened to hapū. Hapū also means sub-tribe. Iwi can refer to bone, people and strength but in this instance means tribe (Williams 1971).
3. Other ways in which Māori women may experience pregnancy differently from Pākehā and other non-Māori women is through involvement with Māori initiatives in the area of health. Since its establishment in 1951, the Māori Women's Welfare League has worked to address the needs of Māori women and their families. The League began to devote more time to labour market issues during the 1970s, but health including family planning and pre- and post-natal care for mothers and babies have been a continuing concern. Amongst the changes to New Zealand's greatly restructured health system over the last decade have been Māori initiatives put in place specifically to improve the health status of Māori. The Tipu Ora Program based in Rotorua which provides health promotion and intervention for Māori mothers and their children, is just one example. Tipu Ora began in November 1991 as a one year pilot study program to increase Māori women's and children's access to, and utilisation of, health services. Another component is the teaching of parenting and life skills to young mothers. The program identifies and addresses the health needs of Māori mothers and their children from conception to when kohanga reo ends, and it delivers services at the most

accessible place for the mother and child (i.e. the marae, kohanga reo or the home) by kaitiaki (caregivers) from the regions (Statistics New Zealand 1993: 154–155).

4. Although I have only discussed Māori and Pākehā customs and practices that pertain to pregnancy and birth, there are of course many other individuals and groups within Aotearoa New Zealand who understand maternities in different ways. See for example Dawson (1983) who interviewed twelve migrant women — including women from various regions in India, Samoa, the Cook Islands, Tonga, China, Laos, Cambodia, Vietnam and England — who all now live in Wellington, Aotearoa New Zealand. They each spoke about their own experiences and customs of pregnancy and birth. Through varying cultural practices each woman was psychically and corporeally inscribed in different ways. The fact that they all now live in the same region — Wellington — does not necessarily lead to shared experiences and understandings of pregnancy, birth and mothering. More recent work on the ways in which new migrant mothers adjust to parenthood in New Zealand has been carried out by Ruth De Souza (2006).
5. Joan Donley passed away on 4 December 2005 at the age of 89. She received an OBE (order of the British Empire), for services to New Zealand and midwifery. She was a feminist and health activist, founder of the New Zealand College of Midwives and the Domiciliary Midwifery Society, and author of a number of books on midwifery and birth including *Birthrites* (1999). Falconer (2005) writes in her obituary 'In her 30-odd years as a midwife in Auckland, Joan Donley delivered, or 'caught' as she preferred to say, between 700 and 800 babies. One of the first arrived on her dining-room table. When she began her career in New Zealand, in the early 1970s, she was one of only three domiciliary, or home-based, midwives in the country. Now the New Zealand College of Midwives, which she helped to found, represents more than 2000 members' (see http://www.nzherald.co.nz)
6. Hanson (1982) discusses tapu and noa in relation to menstruating, pregnant and birthing Māori and Polynesian women although he focuses on historical rather than contemporary understandings of the concepts. Hanson argues that many authors, such as Elsdon Best (1975) and Jean Smith (1974), claim that the atua 'found female genitalia utterly contaminating, repulsive'. Atua would 'retreat before women and anything women contacted, leaving those things devoid of godly influence, which is to say, in a noa state' (Hanson 1982: 346). Hanson (1982: 346–347) refers to this as the 'repellent thesis' and asks if the atua are repelled by women and female genitalia then how is it that pregnant women were (are) considered tapu? Hanson puts forward what he calls an 'affinity thesis'. He argues that the vagina and birth canal are not repellent to the atua but rather they are a conduit between the human and the ultrahuman world of the gods. Women are not pollutant but potent.
7. Some authors have noted that according to 'tradition' Māori women who were pregnant were allowed to eat whatever they liked (see Makereti 1986: 112–113; Riley 1994: 87). If they longed for any kind of food, it was procured for them, no matter how difficult it was to get. Makereti (1986: 112) claims that sometimes this meant expeditions to a distant part of the forest or country, and these expeditions were likely to be very difficult since there were in many cases no roads. Makereti also notes that these foods, although procured for the mother, were really for the child. It was believed that the fetus made decisions as to which foods should be eaten. It ought to be noted, however, that much of this literature on traditional Māori practices surrounding pregnancy and birth relates to women in families of high standing. For such

families there was often much ceremony connected with birth. It is more difficult to find out about the ways in which pregnant women from families with lower standing understood and experienced pregnancy.

8. Tuaiwa (Eva) Rickard (1925–1997) became a well known figure in New Zealand in 1978 when she was arrested with a group of protestors on Raglan Golf Course. The land was taken by the government during the war and not returned to its rightful Māori owners. Mrs Rickard gave birth to nine children. She was passionate about the well being of her children, whanau, hapu and iwi, and was committed to exposing environmental injustices and racist practices in Aotearoa.
9. Figures on household composition from the 2001 Census reveal that in 2001 fewer than half of all households were comprised of a heterosexual couple who live with children (Statistics New Zealand 2001). Given that the majority of housing (both old and new) was designed for nuclear families 'the vast majority of New Zealanders occupy domestic places that may well feel like wearing shoes that do not fit properly' (Longhurst 1999: 157).

CHAPTER 7

1. I sought permission from the New Zealand Ministry of Health to reproduce this poster but was told that it had been embargoed and that this embargo had not been lifted. On the advice of the Women's Health Action Trust I wrote a letter to the Ministry (via the Trust) dated 3 August 2005 requesting a release from the embargo but never received a response.
2. These stories of interspecies breastfeeding remind me of the Roman story of twin brothers Romulus and Remus. Romulus and Remus were the sons of the God Mars. When they were very young they were abandoned by the banks of the River Tiber and left to fend for themselves. Fortunately for them they were discovered by a wolf who fed them with her milk.
3. Some men develop large breasts and their prolactin levels increase to a point that brings on lactation. Others do not have large breasts but attempt to induce lactation. Women who are not pregnant can also induce lactation. This is because the hormones which govern lactation, prolactin and oxytocin, are pituitary, not ovarian hormones. It is possible for a woman, even if she is postmenopausal or had a hysterectomy, to induce lactation in response to nipple stimulation. There are now several regimens which use hormone therapy to assist in bringing in milk, but many women have induced lactation with only mechanical stimulation such as breast massage, nipple manipulation, and sucking - either by a baby or a hospital grade electric breast pump. Some adopting mothers rent a breast pump in anticipation of the infant; other mothers simply put the adopted infant to breast. Giles (2003: 105) explains adoptive mothers can now obtain assistance from lactation consultants, adoptions agencies and Internet sites on how best to induce lactation so they can breastfeed their babies (see Nemba 1994). Many women who induce lactation do not produce enough milk to completely satisfy their baby's needs but even some breast milk is considered useful for the baby's health and for the quality of the mother-child relationship.
4. See also the BBC documentary 'Boys with Breasts', screened in the UK 18 August 2004 which features David and Laura Shanley talking about male lactation.

5. On the 26 February 2003 Ms Kirstie Marshall, a 33 year old Australian Labour Party backbencher of the Victorian Legislative Assembly, was ejected from the chamber for breastfeeding her 12-day-old baby. The reason given was that according to Parliamentary rules only MPs and certain parliamentary staff are allowed in the House during sitting times. The baby was classified as a 'stranger in the House' (non-elected person) and therefore could not be permitted to stay (Wikipedia Kirstie Marshall 2006 <http://en.wikipedia.org/wiki/Kirstie_Marshall> accessed 5 September 2006).

CHAPTER 8

1. From this point on this chapter I do not use inverted comas around the words good and bad but I still wish to acknowledge the problematic nature of these terms. The terms are in common usage but of course what constitutes good and bad is open to contestation. Also the terms do not necessarily exist in a binary relationship.
2. This research was carried out by Amanda Banks for a University of Waikato masters thesis entitled 'Reinforcing and resisting the "DPB mum" as lacking: Pākehā women claiming the DPB in Hamilton, New Zealand'. I supervised the project.

CHAPTER 9

1. Dodge and Kitchen (2001) explain the term cyberspace literally means 'navigable space' and is derived from the Greek word kyber (to navigate). Dodge and Kitchen (2001: 1) use the term cyberspace to refer 'to the conceptual space within ICTs (information and communication technologies), rather than the technology itself. At present, cyberspace does not consist of one homogeneous space; it is a myriad of rapidly expanding cyberspaces, each providing a different form of digital interaction and communication.' Dodge and Kitchen identify three main categories of cyberspace, the Internet (and intranets), virtual reality and more conventional telecommunications such as phone and fax. In this chapter I focus on the first category of cyberspace — the Internet.
2. It is worth pausing to note here that when I conducted interviews with first time pregnant women in Hamilton in 2003 although I did not ask participants about their computer-mediated communications on conception, pregnancy, birth, infant feeding and parenting the topic emerged. For example, one participant explained that after the first couple of months she clicked into a website which told her every week what to expect. When I asked another participant if she had looked at or purchased maternity clothing she replied that she had mostly looked on the Internet. She said she had not actually purchased much because most of the sites are in the US and they won't ship to New Zealand but she had 'checked them out' because they were interesting.
3. The questions Steve Jones encouraged contributors to the collection to grapple with were: Why should we do Internet research? What methods are scholars already using to study the Internet? What methods could we use but have not yet? What are the advantages and disadvantages to these methods and others? What can a particular method, area of interest, mode of inquiry, or way of asking questions contribute to Internet research? (S. Jones 1999: xi–xiv).

These questions were useful in guiding my thinking about how, why and what to search for (more about the methodological process used to conduct research for this chapter follows).

4. Of course it is not just mothers who now have increased access to a range of goods and services that can be purchased online but anyone with ICT access. In this chapter, however, the focus rests on retailing to becoming and new mothers.
5. Hysteria is usually defined today as a mental disorder characterized by emotional outburst, susceptibility to autosuggestion and often physical symptoms such as paralysis (see Grosz 1994: 38–40). The word hysteric is derived from the seventeenth century Latin word *hystericus* literally meaning: of the womb. In the nineteenth century there was a belief that hysteria in women originated in disorders of the womb. More specifically, hysteria in women was thought to be the result of the womb moving around the body and getting too close to the brain. In contemporary accounts of pregnancy still exists evidence that pregnant women are likely to exhibit behaviours such as becoming 'overly' emotional, not being able to think clearly or rationally and frequently forgetting things that prior to pregnancy they would have had no difficulty in remembering or dealing with (Longhurst 1997).
6. There are some interesting links here with consumption practices for people with agoraphobia who are also likely to favour on-line purchasing (see Davidson 2001 and 2003).
7. In 2005 the Maternity Centre Association reconsidered their mission, vision and beliefs and decided that they needed a new name and look to communicate their contemporary focus. They decided upon Childbirth Connection, and tagline, 'Helping women and health professionals make informed maternity care decisions.' Their mission remains the same, that is, 'to promote safe, effective and satisfying care for all women and their families through research, education and advocacy' (Childbirth Connection, available http://www.childbirthconnection.org/, accessed 11 October 2006).

CHAPTER 10

1. 'Unions of Fathers' or 'Fathers' Unions' have been set up in many countries including the United Kingdom, United States, Canada, Australia and New Zealand by men who feel that fathers are treated unfairly by family courts and child welfare agencies in custody disputes. They argue children have a right to a life that includes both a father and a mother when a relationship breaks up.

Bibliography

A healthy me: forgetfulness (2000) Available <http://www.ahealthyme.com/article/bcrefcap/100322064> (accessed 10 August 2005).

Abel, S. and Kearns, R.A. (1991) 'Birth places: a geographical perspective on planned home birth in New Zealand', *Social Science & Medicine*, 33, 7: 825–834.

Adams, P. (1997) 'Cyberspace and virtual places', *The Geographical Review*, 87, 2: 155–71.

——— (1998) 'Network typologies and virtual place', *Annals of the Association of American Geographers*, 88, 1: 88–106.

Adams, P.C. and Ghose, R. (2003) 'India.com: the construction of a space between', *Progress in Human Geography*, 27, 4: 414–37.

Aitken, S.C. (2000) 'Fathering and faltering: "Sorry, but you don't have the necessary accoutrements"', *Environment and Planning A*, 32: 581–598.

——— (2005) 'Textual analysis: reading culture and context', in R. Flowerdew and D. Martin (eds), 2nd edition, *Methods in Human Geography: a guide for students doing a research project*. Edinburgh Gate, Pearson Education: pp. 233–249.

Ananova (2005) Available <http://www.ananova.com/news/story/sm_700634.html> (accessed 30 May 2005).

August, W. (2004) 'The Māori female — her body, spirituality, sacredness and mana: a space within spaces', unpublished thesis, University of Waikato.

——— (2005) Māori women: bodies, spaces, sacredness and mana, *New Zealand Geographer*, 61, 2: 117–123.

Babyworld (2005) Available <http://www.babyworld.co.uk/shop> (accessed 10 July 2005).

Bad Mothers Club (2006) Available <http://www.badmothersclub.co.uk/jsp/index.jsp> (accessed 12 September 2006).

Bailey, L. (1999) 'Refracted selves? A study of changes in self-identity in the transition to motherhood', *Sociology*, 33, 2: 335–52.

——— (2001) 'Gender shows: first-time mothers and embodied selves' *Gender and Society*, 15, 1: 110–129.

Bailey, R. (1992) 'Clothes encounters of the gynecological kind: medical mandates and maternity modes in the USA, 1850–1990', in R. Barnes and J.B. Eicher (eds) *Dress and Gender: making and meaning*, pp. 248–65, New York: Berg.

Balaskas, J. (1989) *The New Active Birth Handbook*, New York: Unwin Hyman.

Banim, M., Green, E. and Guy, A. (2001) Introduction, in A. Guy, E. Green and M. Banim (eds) *Through the Wardrobe: women's relationships with their clothes*, pp. 1–17, Oxford: Berg.

Banks, A. (2005) 'Reinforcing and resisting the 'DPB mum' as lacking: Pākehā women claiming the DPB in Hamilton, New Zealand', unpublished thesis, University of Waikato.

Banks, M. (2000) *Home Birth Bound: mending the broken weave*, Hamilton, New Zealand: Birthspirit Books.

Bartlett, A. (2004) 'Black breasts, white milk? ways of constructing breastfeeding and race in Australia', *Australian Feminist Studies*, 19, 45: 341–55.

Beale, V. (2003) Fashioning the Pregnant Body: wearing pregnant bodies, unpublished PhD thesis, University of Sheffield.

Bears from Robin Rive (2006) Available <http://www.bearshops.co.uk/manuf04.html> (accessed 12 September 2006).

Beere, P. and Brabyn, L. (2006) 'Providing the evidence: Geographic accessibility of maternity units in New Zealand', *New Zealand Geographer*, 62, 2: 135–143.

Bell, D., Binnie, J., Cream, J. and Valentine, G. (1994) 'All hyped up and no place to go', *Gender, Place and Culture: A Journal of Feminist Geography*, 1, 1: 31–47.

Bell, D. and Valentine, G. (1995) 'Introduction: Orientations', in D. Bell and G. Valentine (eds) *Mapping Desire: geographies of sexualities*, London: Routledge, 1–27.

—— (1997) *Consuming Geographies: we are where we eat*, London: Routledge.

Best, E. (1924) *The Maori, Memoirs of the Polynesian Society*, Wellington: Board of Maori Ethnological Research.

—— (1975) (originally published 1929) *The Whare Kohanga (the 'Nest House') and its Lore*, Wellington: A.R. Shearer.

Better Living Discussion Forum (2004) Available <http://www.betterliving.com.nz/net/forums/> (accessed 3 December 2004).

Bhabha, H. K. (1994): *The Location of Culture*, London: Routledge.

Binnie, J. and Valentine, G. (1999) 'Geographies of sexualities — a review of progress', *Progress in Human Geography*, 23, 2: 175–87.

Blunt, A. and Dowling, R. (2006) *Home*, New York: Routledge.

Bondi, L. and Davidson, J. (2005) 'Situating gender', in L. Nelson and J. Seager (eds) *A Companion to Feminist Geography*, 15–31, Malden: Blackwell.

Bonell, C. (2004) 'Why is teenage pregnancy conceptualized as a social problem? A review of quantitative research from the USA and UK', *Culture, Health and Sexuality*, 6, 3: 255–272.

Boots (2006) Available <http://www.boots.com/> (accessed 2 October 2006).

Bordo, S. (1990) 'Reading the slender body', in M. Jacobus, E. Fox Keller and S. Shuttleworth (eds), *Body/Politics: women and the discourses of science*, 83–112, New York: Routledge.

—— (1993) *Unbearable Weight: feminism, Western culture and the body*, London: University of California Press.

Bornfree (2005) Available <www.unassistedchildbirth.com> (accessed 3 June 2005).

Bourdieu, P. (1990). *The Logic of Practice*, Stanford: Stanford University Press.

'Boys with breasts' (screened 18 April 2004) BBC documentary screened Channel 4, United Kingdom.

Braziel, J.E. and LeBesco, K. (eds) (2001) *Bodies Out of Bounds: fatness and transgression*, Berkeley: University of California Press.

Brown, M. A. (1987) Employment during pregnancy: Influences on Women's Health and Social Support, in K.M. Swanson-Kauffman, *Women's Work, Families, and Health. The balancing act*, 51–67, New York: Hemisphere.

Brown, M. P. (2000) *Closet Space: geographies of metaphor from the body to the globe*, London: Routledge.

Brown, M. and Knopp, L. (2003) 'Queer cultural geographies — We're here! We're queer! We're over there, too!', in K. Anderson, M. Domosh, S. Pile and N. Thrift (eds) *Handbook of Cultural Geography*, 313–24, London: Sage.

Bryder, L. (2003) *A Voice for Mothers: the Plunket Society and infant welfare 1907–2000*, Auckland: Auckland University Press.

Burns, D. (2000) 'Feminism, Psychology and Social Policy: Constructing political boundaries at the grassroots', *Feminism and Psychology*, 10, 3, 367–380.

Butler, J. (1990) *Gender Trouble: feminism and the subversion of identity*, New York: Routledge.

Butler, R. and Bowlby, S. (1997) 'Bodies and spaces: an exploration of disabled people's experiences of public space', *Environment and Planning D: Society and Space*, 15, 366–368.

Butler, R. and Parr, H. (eds) (1999) *Mind and Body Spaces: geographies of illness, impairment, and disability*, New York: Routledge.

Calman, S. (no date) *Confessions of a Bad Mother: in the aisle by the chill cabinet no-one can hear you scream* (no publisher)

Chapkiss, W. (1986) *Beauty Secrets: women and politics of appearance*, Boston: South End Press.

Chase, S.E. and Rogers, M.F. (2001) *Mothers and Children: feminist analyses and personal narratives*, New Brunswick, N.J: Rutgers University Press.

Childbirth Connection (2006) Available <http://www.childbirthconnection.org/> (accessed 11 October 2006).

Chouinard, V. (1997) 'Making space for disabled differences: challenging ableist geographies', *Environment and Planning D: Society and Space*, 15, 379–387.

Chouinard, V. and Grant, A. (1995) 'On not being anywhere near the "project"', *Antipode*, 27, 137–66.

Christians vow to break smacking law (9 June 2006) Available <http://www.nzherald.co.nz> (accessed 29 August 2006).

Church Gibson, P. (ed) (2004) *More Dirty Looks: gender, pornography and power*, London: British Film Industry.

Clarke, A. and Miller, D. (2002) 'Fashion and anxiety', *Fashion Theory: the Journal of Dress, Body and Culture*, 6, 191–214.

Clarke, V. (2000) 'Lesbian mothers: sameness and difference', *Feminism and Psychology*, 10, 2: 273–278.

Clearblue (2006) Available <http://www.clearplan.com/TellingYourBossYou'rePre gnant.cfm> (accessed 22 February 2006).

Cline, S. (1990) *Just Deserts. Women and food*, London: André Deutsch.

ClubMom (2004) Available <http://www.clubmom.com> (accessed 11 July 2004).

Collins English Dictionary (1979) Glasgow, William Collins Sons & Co.

Collins, H. (1994) *The EU Pregnancy Directive: a guide for human resource managers*, Oxford: Blackwell.

Collins, S. (1995) 'A Pregnant Woman Worker: A Contradiction in Terms?', M. Phil Thesis, University of Auckland.

Consumers' Institute (1994) 'Your choices in childbirth', *Consumer*, March, 324, 6–9.

Craik, J. (1994) *The Face of Fashion: cultural studies in fashion*, London: Routledge.

Crang, M., Crang, P. and May, J. (1999) *Virtual Geographies: bodies, spaces and relations*, London: Routledge.

Cream, J. (1995) 'Women on trial — a private pillory?' in S. Pile and N. Thrift (eds) *Mapping the Subject: geographies of cultural transformation*, 158–69, London: Routledge.

Cresswell, T. (1996) *In Place/Out of Place: geography, ideology and transgression.* Minneapolis: University of Minnesota Press.

——— (1997) 'Weeds, plagues and bodily secretions: a geographical interpretation of metaphors of displacement', *Annals of the Association of American Geographers*, 87, 2: 330–345.

——— (2005) 'Moral geographies', in, D. Atkinson, P. Jackson, D. Sibley and N. Washbourne (eds) *Cultural Geography: a critical dictionary of key concepts*, pp. 128–134, London and New York: Taurus.

Crewe, L. and Davenport, E. (1992) 'The puppet-show: buyer-supplier relations in clothing retailing', *Transactions Institute of British Geographers*, 17, 183–197.

Crewe, L. and Forster, Z. (1993) 'Markets, design and local agglomeration: the role of the small independent retailer and the workings of the fashion system', *Environment and Planning D: Society and Space*, 11, 213–229.

Crewe, L. and Lowe, M. (1996) 'Gap on the map: towards a geography of consumption and identity', *Environment and Planning A*, 29, 1877–1898.

Crooks, V.A. (2006) '"I go on the internet; I always, you know, check to see what's new": Chronically ill women's use of online health information to shape and inform doctor-patient interactions in the space of care provision', *ACME An International E-Journal for Critical Geographies*, Special Issue on Gender, Space and Technology, 5, 1, 50–69.

Cutchin, M. (2002) 'Virtual medical geographies: conceptualizing telemedicine and regionalization', *Progress in Human Geography*, 26, 1: 19–39.

Daly, G. (1996). *Homeless: policies, strategies, and lives on the street*, London: Routledge.

Davidson, J. (2001) 'Fear and trembling in the mall: women, agoraphobia, and body boundaries', in, I. Dyck, N. Davis Lewis, and McLafferty, S. (eds) *Geographies of Women's Health*, New York: Routledge.

——— (2003) *Phobic Geographies: the phenomenology and spatiality of identity*, Aldershot: Ashgate.

Davidson, J. Bondi, L. and Smith, M. (eds) (2005) *Emotional Geographies*, Aldershot: Ashgate.

Davis, K. (1997) 'Embody-ing theory', in K. Davis (ed.) *Embodied Practices: feminist perspectives on the body*, 1–23, London: Sage.

Davis, T. (1995) 'The diversity of queer politics and the redefinition of sexual identity and community in urban spaces', in D. Bell and G. Valentine (eds) *Mapping Desire*, 284–303, London: Routledge.

Davis-Floyd, R. (1992) *Birth as an American Rite of Passage*, Berkeley: University of California Press.

——— (1994) 'Mind over body: the pregnant professional', in N. Sault (ed) *Many Mirrors: body image and social relations*, 204–233, New Jersey: Rutgers University Press.

Dawson, R. 1983: *Customs of Childbirth: migrant women from twelve different cultures speak of their own experiences and customs*, Wellington Multicultural Educational Resource Centre.

de La Haye, A. and Wilson, E. (eds) (1999) *Defining Dress: dress as object, meaning and identity*, Manchester: Manchester University Press.

Deem, H. and N.P. Fitzgibbon (1953) *Modern Mothercraft: a guide to parents*, Official Handbook Royal New Zealand Society for the Health of Women and Children, Dunedin: (Incorporated) Plunket Society.

Derrida, J. (1974) *Of Grammatology*, trans. G.C. Spivak, Baltimore: John Hopkins University Press.

—— (1978) *Writing and Difference*, trans. A. Bass, Chicago: University of Chicago Press.

—— (1981) *Dissemination*, trans. B. Johnson, Chicago: University of Chicago Press.

DeSouza, R. (2006). New spaces and possibilities: The adjustment to parenthood for new migrant mothers. Wellington: Families Commission. Available <http://www.familiescommission.govt.nz/publications/blue-skies.php> (accessed 23 January 2007).

DiQuinzio, P. (1999) *The Impossibility of Motherhood: feminism, individualism, and the problem of mothering*, New York: Routledge.

Disability, pregnancy & parenthood international (2005) Available <http://www.dppi.org.uk/> (accessed 27 July 2005).

Dodge, M. and Kitchen, R. (2001) *Mapping Cyberspace*, London: Routledge.

Donley, J. (1986) *Birthrites. Natural vs unnatural childbirth in New Zealand*, Auckland, The Full Court Press, in association with NZ College of Midwives, Christchurch.

—— (1992) *Herstory of NZ Homebirth Association*, The New Zealand Homebirth Association: Auckland.

Doucet, A. (2006) *Do Men Mother? Fathering, care, and domestic responsibility*, Toronto: University of Toronto Press.

Douglas, M. (1966) *Purity and Danger*, London: Routledge and Kegan Paul.

Draper, J. (2003) 'Blurring, moving and broken boundaries: men's encounters with the pregnant body', *Sociology of Health & Illness*, 25, 7: 734–767.

Driver, F. (1988) 'Moral geographies: social science and the urban environment in mid-nineteenth-century England', *Transactions of the Institute of British Geographers*, 13: 275–278.

Duncan, N. (ed) (1996) *BodySpace: destabilizing geographies of gender and sexuality*, London: Routledge.

Dwyer, C. (1999) 'Veiled meanings: Young British Muslim women and the negotiation of differences', *Gender, Place and Culture*, 6, 1, pp. 5–26.

Dyck, I. (1995) 'Hidden geographies: the changing lifeworlds of women with multiple sclerosis', *Social Science & Medicine*, 40, 3: 307–20.

—— (1999) 'Body troubles: women, the workplace and negotiations of a disabled identity', Butler, R. and Parr, H. (eds) *Mind and Body Spaces: geographies of illness, impairment and disability*, London: Routledge, 119–137.

Dyck, I. Davis Lewis, N. and S. McLafferty (2001) (eds) *Geographies of Women's Health*, New York: Routledge.

Earle, S. (2003) "Bumps and boobs': fatness and women's experiences of pregnancy', *Women's Studies International Forum*, 26, pp. 245–252.

Ekinsmyth, C., Elmhirst, R., Holloway, S. and Jarvis, H. (2004) 'Love changes all: making some noise by 'coming out' as mothers', *WGSG, Geography and Gender Reconsidered*, 95–107.

Elder, G. (1999) "Queerying' boundaries in the geography classroom', *Journal of Geography in Higher Education*, 23, 1: 86–93.

Ettorre, E. (2000) 'Reproductive Genetics, Gender and the Body: 'Please Doctor, may I have a Normal Baby?'', *Sociology*, 34, 3: 403–20.

Falconer, P. (2005) Obituary for Joan Donley, *New Zealand Herald*, 17 December. Available <nzhearld.co.nz> (assessed 24 August 2006).

Families South East. Join the Bad Mothers Club (2006) Available <http://www.familiesonline.co.uk/article/articleview/443/1/45> (accessed 7 September 2006).

Fannin, M. (2003) 'Domesticating birth in the hospital: "family centered" birth and the emergence of "homelike" birthing rooms', *Antipode*, 34, 3: 513–535.

——— (2006) 'Global midwifery and the technologies of emotion, *ACME An International E-Journal for Critical Geographies*, Special Issue on Gender, Space and Technology, 5, 1, 70–88

Firestone, S. 1971: *The Dialectic of Sex: the case for feminist revolution,* London, Cape.

Fluri, J.L. (2006) '"Our website was revolutionary" Virtual spaces of representation and resistance', *ACME An International E-Journal for Critical Geographies*, Special Issue on Gender, Space and Technology, 5, 1, 89–111.

Foucault, M. (1977) *Discipline and Punish: the birth of the prison*, London: Penguin.

——— (1980) *Power/Knowledge*, New York: Pantheon.

Fuss, D. (ed) (1991) *Inside/Out: lesbian theories, gay theories*, New York: Routledge.

Gabb, J. (2004) "I could eat my baby to bits'; passion and desire in lesbian mother-children love', *Gender, Place and Culture*, 11, 3: 399–415.

Garland-Thomson, R. (2005) 'Feminist disability studies', *Signs: Journal of Women in Culture and Society*, 30, 2, 1557–1587.

Garner, T. (2003) 'Breastfeeding poster dropped', *Weekend Herald*, A3.

Gatens, M. (1988) 'Towards a feminist philosophy of the body', in B. Caine, E. Grosz and M. de Lepervanche (eds) *Crossing Boundaries: feminisms and critiques of knowledges*, 59–70, Sydney: Allen and Unwin.

——— (1991) 'Corporeal representation in/and the body politic', in R. Diprose and R. Ferrell (eds) *Cartographies: poststructuralism and the mapping of bodies and spaces*, 79–87, Sydney: Allen and Unwin.

Giles, F. (2004) '"Relational, and Strange": a Preliminary Foray into a Project to Queer Breastfeeding', *Australian Feminist Studies*, 19, 45, 302–315.

——— (2003) *Fresh Milk: the secret life of breasts*, New York: Simon & Schuster

Gill, C.J. (1996) 'Becoming visible: Personal health experiences of women with disabilities', in D.M. Krotoski, M.A. Nosek, & M.A. Turk (eds) *Women with Physical Disabilities: achieving and maintaining health and well-being*, 5–15, Baltimore: Raul H. Brooks Publishing Co.

——— (1997) 'The last sisters: health issues of women with disabilities', in S.B. Ruzek, V.L. Olesen and A.E. Clarke (eds) *Women's Health: complexities and differences*, Columbus, OH, Ohio State University Press, 96–111.

Girl.Mom. Support, Community, and Education for Young Mamas (2006) Available <http://www.girl-mom.com/index.php> (accessed 18 April 2006).

Gleeson, B. (1999) *Geographies of Disability*, New York: Routledge.

Goffman, E. (1968) *Stigma: notes of the management of spoiled identity*, Harmondsworth: Penguin.

Green, E. (2001) 'Suiting ourselves: women professors using clothes to signal authority, belonging and personal style', in: A. Guy, E. Green and M. Banim (eds) *Through the Wardrobe: women's relationships with their clothes*, pp. 203–220, Oxford: Berg.

Gregson, N. and Beale, V. (2004) 'Wardrobe matter: the sorting, displacement and circulation of women's clothing', *Geoforum*. Themed Section on 'Material Geographies', 35, 6, 689–700.

Gregson, N., Brooks, K and Crewe, L. (2001) 'Bjorn Again? Rethinking 70s revivalism through the reappropriation of 70s clothing', *Fashion Theory*, 5, 3, 3–28.

Gregson, N. and Rose, G. (2000) 'Taking Butler elsewhere: performativities, spatialities and subjectivities', *Environment and Planning D: Society and Space*, 18, 4: 433–452.

Grosz, E. (1989) *Sexual Subversions: three french feminists*, Sydney: Allen and Unwin.

—— (1994) *Volatile Bodies: toward a corporeal feminism*, St Leonards: Allen and Unwin.

Grue, L. and Tafjord Lærum, K. (2002) '"Doing motherhood": some experiences of mothers with physical disabilities', *Disability and Society*, 17, 6: 671–83.

Gurak, L.J. and Ebeltoft-Kraske, L. (1999) 'The rhetorics of gender in computer-mediated communication', *The Information Society*, 15, 3: 147–50.

Guy, A. (2001) 'Dis/continued selves: why do women keep clothes they no longer wear?' in Ali. Guy, Eileen Green and Maura Banim (eds) *Through the Wardrobe: women's relationships with their clothes*, pp. 203–220, Oxford: Berg.

Guy, C. 1992: Maternity leave — pregnant pause, *New Zealand Herald*, September 30, 33.

Haar, J. and Dyer, S. (2000) 'Redirecting Work-Family Research: The argument for specific pregnancy programmes', *Department of Strategic Management and Leadership Working Paper Series 2*. University of Waikato, Hamilton, New Zealand.

Hanson, F.A. (1982) 'Female pollution in Polynesia?' *Journal of Polynesian Society*, 91, 1: 335–381.

Hart, A. (1995) '(Re)constructing a Spanish red-light district: prostitution, space and power', in D. Bell, G. Valentine (eds) *Mapping Desire: geographies of sexualities*, pp. 214–228, London: Routledge.

Hillis, K. (1996) 'A geography of the eye: the technologies of virtual reality', in R. Shields (ed) *Cultures of Internet: virtual spaces, real histories, living bodies*, 70–98, London: Sage.

Hodge, K. (18 November 2004) 'Breastfed dog may turn savage, say experts', *The Dominion Post*. Available <http://io.knowledge-basket.co.nz> (accessed 31 May 2005).

—— (19 November 2004) Treatment for woman who breastfed puppy. *The Dominion Post*. Available <http://io.knowledge-basket.co.nz> (accessed 31 May 2005).

—— (20 November 2004) Puppy woman loses child. The Dominion Post. Available <http://io.knowledge-basket.co.nz> (accessed 31 May 2005).

Holloway, S.L. (1998a) 'Geographies of justice: preschool-children provision and the conceptualization of social justice', *Environment and Planning C: Government and Policy*, 16, 1: 85–104.

—— (1998b) 'Local childcare cultures: moral geographies of mothering and the social organization of pre-school education', *Gender, Place and Culture*, 5, 1: 29–53.

—— (1998c) "She lets me go out once a week': mothers' strategies for obtaining personal time and space', *Area*, 30, 4: 321–30.

—— (1999) 'Reproducing motherhood', in Laurie, N., Dwyer, C., Holloway, S. and Smith, F., *Geographies of New Femininities*, 91–112, Longman: Edinburgh Gate.

Holloway, S.L., and Valentine, G. (2001a) "It's only as stupid as you are': children's and adults' negotiation of ICT competence at home and at school', *Social and Cultural Geography*, 2, 1: 25–42.

—— (2001b) 'Children at home in the wired world: reshaping and rethinking the home in urban geography', *Urban Geography*, 22, 6: 562–83.

—— (2001c) 'Placing cyberspace: processes of Americanization in British children's use of the Internet', *Area*, 33. 2: 153–60.

Holmes, P. (screened 18 July 2003) Current Affairs Programme. Interview with Brooke — a prostitute from Hawera, New Zealand who is offering breast-milk to clients, Television One, New Zealand.
Holt, M. (12 October 2002) 'Porn', *Waikato Times*, edition 2, 1.
——— (14 October 2002) 'Deal struck to film porn star giving birth', *Waikato Times*, edition 2, 2.
——— (19 October 2002) 'Birth rites', *Waikato Times*, edition 2, 19.
——— (24 October 2002) 'Scrapped porn birth stuns star', *Waikato Times*, edition 2, 1.
——— (26 October 2002) 'Porn star speaks out over birth film ban', *Waikato Times*, edition 2, 3.
——— (30 October 2002) 'It's a girl! Porn actress now a mum', *Waikato Times*, edition 2, 1.
hooks, b. (1991) 'Homeplace: a site of resistance', excerpts from *Yearning: race, gender and cultural politics*, Boston: South End Press; in L. McDowell (ed) *Undoing Place? A geographical reader*, 33–38, London: Arnold.
Hubbard, P. (1997) 'Red-light districts and toleration zones: geographies of female street prostitution in England and Wales', *Area*, 29, 2: 129–40.
——— (1998) 'Sexuality, immorality and the city: red-light districts and the marginalization of female street prostitutes', *Gender, Place and Culture*, 5, 1: 55–74.
——— (2000) 'Desire/disgust: mapping the moral contours of heterosexuality', *Progress in Human Geography*, 24, 2: 191–217.
Huggies (2006) Available <http://www.huggiesclub.com/> Accessed 2 October 2006.
Huntly. Welcome to our town (2005) Available <http://www.huntly.net.nz/> (accessed 23 March 2005).
Irigaray, L. (1985a) *This Sex Which is Not One*, trans. C. Porter with C. Burke, New York: Cornell University Press.
——— (1985b) *Speculum of the Other Woman*, trans. G.C. Gill, New York: Cornell University Press.
Irmie, R. (1996) *Disability and the City: international perspectives*, New York: St Martin's Press.
Irving, D. (30 May 2005) 'Breast is best for Matilda', *Waikato Times*, 7.
iVillage.co.uk 2006 (2006) Available <http://www.ivillage.co.uk/pregnancyandbaby> (accessed 2 October 2006).
Janelle, D. and Hodge, D. (eds) (2000) *Information, Place, and Cyberspace: issues in accessibility*, Berlin: Springer-Verlag.
Johnson, L. (1989) 'Embodying geography — some implications of considering the sexed body in space', New Zealand Geographical Society Proceedings of the 15th New Zealand Geography Conference, Dunedin, New Zealand, August, 134–38.
Johnson, L., with Huggins, J. and Jacobs, J. (2000) *Placebound. Australian feminist geographies*, Melbourne: Oxford University Press.
Johnson, R. (2004) 'Bars, breasts, babies: Justice L'Heureux-Dubé and the boundaries of belonging', in E. Sheehy (ed.) *Adding Feminism to the Law: the contributions of Justice Claire L'Heureux-Dubé*, Irwin Law, 139–57.
——— (2005) 'Law and the leaky woman: the saloon, the liquor license, and narratives of containment', *Continuum: journal of media and cultural studies*, 19, 2: 181–99.
Johnston, L. (1997) 'Queen(s') street or Ponsonby poofters? the embodied HERO parade sites', *New Zealand Geographer*, 53, 2: 29–33.
——— (2005) *Queering Tourism: paradoxical performances at gay pride parades*, London, Routledge.

—— (2007) Mobilising Pride/Shame: Lesbians, tourism and parades, *Social and Cultural Geography,* 8, 1: 29–45.
Johnston, L. and Valentine, G. (1995) 'Wherever I lay my girlfriend, that's my home: the performance and surveillance of lesbian identities in domestic environments', in D. Bell and G. Valentine (eds) *Mapping Desire: geographies of sexualities*, 99–113, London: Routledge.
Johnston, R.J., Gregory, D., Pratt, G. and Watts, M. (eds) (2000) *The Dictionary of Human Geography*, Massachusetts: Blackwell.
Jolly, M. (1998) 'Colonial and post-colonial plots in histories of maternities and modernities', in Ram, K. and Jolly, M. (eds) *Maternities and Modernities: colonial and post-colonial experiences in Asia and the Pacific*, Cambridge: Cambridge University Press, 1–25.
Jones, D. (1999) 'Pop Belly', Sunday Times 28 February, pp. 32–40. Cited in Tyler, I. (2001) 'Celebrity, pregnancy and subjectivity', in S. Ahmed and J. Stacey (eds) *Thinking Through the Skin*, 69–83, London: Routledge.
Jones, S. (ed) (1999) *Doing Internet Research: critical issues and methods for examining the Net*, Thousand Oaks: Sage.
Katz, C. and Monk, J. (eds) (1993) *Full Circles. Geographies of women over the life course*, London: Routledge.
Katz Rothman, B. (1982) *In Labour: women and power in the birthplace*, London: Junction Books.
Kellner, D. (1994) 'Madonna, fashion, and identity', in: S. Benstock and S. Ferriss (eds) *On Fashion*, pp. 159–182, New Brunswick: Rutgers University Press.
Kenny, J.W. and Tash, D.T. (1993) 'Lesbian childbearing co dilemmas and decisions', in Noerager Stern, P. (ed) *Lesbian Health: What are the issues?*, Washington D.C.: Taylor and Francis: 119–129.
Kent, D. (2002) 'Beyond expectations: being blind and becoming a mother', *Sexuality and Disability*, 20, 1: 81–88.
King, F.T. (1913) *The New Zealand Scheme for Promoting the Health of Women and Children*, Reprint of an Address given at the National Congress, London: August 1913.
King intervenes over porn birth (14 October 2002) Available http://onenews.nzoom.com/onenews (accessed 16 February 2005).
Kingfisher, C.P. (1996) *Women in the American Welfare Trap*, Philadelphia: University of Pennsylvania Press.
—— (1999) 'Rhetoric of (female) savagery: welfare reform in the United States and Aotearoa/New Zealand', *National Women's Studies Association Journal*, 11, 1: 1–20.
—— (ed) (2002) *Western Welfare in Decline: globalization and women's poverty*, Philadelphia: University of Pennsylvania Press.
Kitchen, R. (1998a) 'Towards geographies of cyberspace', *Progress in Human Geography*, 22, 3: 385–406.
—— (1998b) 'Out of place', 'knowing one's place': Space, power, and the exclusion of disabled people', *Disability and Society*, 13, 3, 343–356.
Kitzinger, S. (1989) *Pregnancy and Childbirth*, Sydney: Doubleday.
—— (1995) *Ourselves as Mothers: the universal experience of motherhood*, Massachusetts: Addison–Wesley.
Klein, H. (1991) 'Couvade syndrome: male counterpart to pregnancy', *International Journal of Psychiatry Medicine*, 21, 1: 57–69.
Klein, R.S. (1990) *Object Relations and the Family Process*, New York, Praeger.
Kristeva, J. (1980) *Desire in Language: a semiotic approach to literature and art*, Oxford: Basil Blackwell.
—— (1981) 'Women's time', *Signs*, trans. A. Jardine and H. Blake, 7, 1: 13–35.

——— (1982) *Powers of Horror: an essay on abjection*, trans. L.S. Roudiez, New York: Columbia University Press.

Ladd-Taylor, M. and Umansky, L. (1998) *'Bad' Mothers: the politics of blame in twentieth-century America,* New York: New York University Press.

Larner, W. and Spoonley, P. (1995) 'Postcolonial politics in Aotearoa/New Zealand', in D. Stasiulis and N. Yuval-Davis (eds) *Unsettling Settler Societies: articulations of gender, race, ethnicity and class*, London: Sage, pp.39–64.

Leboyer, F. 1975: *Birth Without Violence*, New York: Random House.

Leder, D. (1990) *The Absent Body*, Chicago: University of Chicago Press.

Lockett, C. (2001) 'Warrior princess in stunning new role', *Women's Health Update Magazine*, 6, 3. Available <http://www.lucylawless.info/articles/womenshealth/> (accessed 21 July 2005).

Long, K. (2005) *The Bad Mother's Handbook*, Penguin: London.

Longhurst, R. (1997) 'Going nuts': re-presenting pregnant women, *New Zealand Geographer*, 53, 2: 34–39.

——— (1999) 'Pregnant bodies, public scrutiny: 'giving' advice to pregnant women', in E.K. Teather (ed.) *Embodied Geographies: spaces, bodies and rites of passage*, 78–90, London: Routledge.

——— (2000) ''Corporeographies' of pregnancy: 'bikini babes'', *Environment and Planning D: Society and Space* 18, 4: 453–72.

——— (2001a) 'Trim, taut, terrific, and pregnant', in D. Bell, J. Binnie, R. Holliday, R. Longhurst and R. Peace *Pleasure Zones: bodies, cities, spaces*, 1–28, New York: University of Syracuse Press.

——— (2001b) *Bodies: exploring fluid boundaries*, London: Routledge.

——— (2005a) '(Ad)dressing pregnant bodies in New Zealand: clothing, fashion, subjectivities and spatialities', *Gender, Place and Culture* 12, 4: 433–446.

——— (2005b) 'Man Breasts': spaces of sexual difference, fluidity and abjection, in B. Van Hoven and K. Hoerschelmann (eds) *Spaces of Masculinity*, Routledge: London, pp. 165–78.

Longhurst, R. and Banks, A. (2004) ''Pregnancy chic': pregnant bodies in public places ten years on', in Cortesi, G., Cristaldi, F. and Droogleever Fortuijn, J. (eds) *Gendered Cities: identities, activities, networks: a life-course approach*, 43–56, Rome: Società Geografica Italiana,.

LSR Maternity Collection (2003) Available <http://www.lsrmaternity.com> (accessed 20 November 2003).

MacClancy, J. (1993) *Consuming Culture: why you eat what you eat*, New York: Henry Holt.

Madge, C. and O'Connor, H. (2002) On-line with e-mums: exploring the internet as a medium for research, *Area*, 34: 92–102.

——— (2005) 'Mothers in the making? exploring liminality in cyber/space', *Transactions of the Institute of British Geographers*, 30, 1: 83–97.

——— (2006) Parenting gone wired: empowerment of new mothers on the internet? *Social and Cultural Geography*, 7, 2: 199–220.

Madge, C., Noxolo, P. and Raghuram, P. (2004) 'Bodily contours: geography, metaphor and pregnancy', *WGSG, Geography and Gender Reconsidered*, 68–83.

Mahon-Daly, P. and Andrews, G.J. (2002) 'Liminality and breastfeeding: women negotiating space and two bodies', *Health & Place*, 8, 2: 61–76.

Makereti (1986) (originally published in 1938) *The Old-Time Maori*, Auckland: New Women's Press.

Manderson, L. (1998) 'Shaping reproduction: maternity in early twentieth-century Malaya', in Ram, K. and Jolly, M. (eds) *Maternities and Modernities: colonial and postcolonial experiences in Asia and the Pacific*, 26–49, Cambridge: Cambridge University Press.

Marieskind, H. (1973) Abnormal lactation, *A Journal of Tropical Pediatrics*, 19, 2, 123–128.

Massey, D. (1994) *Space, Place and Gender*, Cambridge: Polity Press.

——— (1996) 'Masculinity, dualisms and high technology', *Transactions of the Institute of British Geographers*, 20, 4: 487–99.

Maternity Clothing Fashions (2006) Available <htpp://www.maternity-clothing-fashions.com> (accessed 2 November 2006).

Maternity Fashion Show (2003) Available <htpp://www/uclh.org/news/bumps0401.shtml> (accessed 4 November 2003).

Maternity Information Booklet, River Ridge East Birth Centre (2006) Available <http://www.riverridgeeastbc.co.nz/index.htm> (accessed 18 April 2006).

Maternity Thong (2003) Available <http://www.newport-news.com/shop/Product.asp> (accessed 20 November 2003).

Matless, D. (1994) 'Moral geography in Broadland', *Ecumene*, 1, 2: 127–55.

Matrix (1984) *Making Space: women and the man made environment*, London: Pluto Press.

McClintock, A. (1995) *Imperial Leather: race, gender, and sexuality in the colonial conquest*, Routledge: New York.

McDowell, L. (1995) 'Body work: heterosexual gender performances in city workplaces', in D. Bell and G. Valentine (eds) *Mapping Desire*, 75–95, Routledge: London.

——— (1997) Capital Culture: gender at work in the city, Oxford: Blackwell.

McDowell, L. and Court, G. (1994) 'Performing work: bodily representations in merchant banks', *Environment and Planning D: Society and Space*, 12, 6: 727–50.

McDowell, L. and Sharp, J.P. (eds) (1999) *A Feminist Glossary of Human Geography*, Arnold: London.

McKeever, P., Angus, J., Lee-Miller, K. and Reid, D. (2003) "It's more of a production': accomplishing mothering using a mobility device', *Disability & Society*, 18, 2: 179–197.

McLeod, R. (2002) 'Children have a right to protection', *Waikato Times*, 23 October 2002.

McLoughlin, D. (1993) 'The politics of childbirth: midwives versus doctors', *North and South*, August: 54–71.

Miller, D., Jackson, P., Thrift, N., Holbrook, B. and Rowlands, M. (1998) *Shopping, Place and Identity*, London: Routledge.

Monk, J. (2001) Many roads: The personal and professional lives of women geographers, in P. Moss (ed) *Placing Autobiography in Geography*, Syracuse, Syracuse University Press, pp. 167–187.

Morgan, D. H. J. and S. Scott (1993) Bodies in a social landscape, in S. Scott and D. H. J. Morgan (eds) *Body Matters*, London: The Falmer Press, 1–21.

Mort, F. (1996) *Cultures of Consumption: masculinities and social space in late twentieth-century Britain*, London: Routledge.

Moss, P. (2001) 'Writing one's life', in P. Moss (ed) *Placing Autobiography in Geography*, pp. 1–21 Syracuse: Syracuse University Press.

Moss, P. and Dyck, I. (1996) Inquiry into environment and body: women, work and chronic illness, *Environment and Planning D: Society and Space*, 14, 737–753.

—— (1999) 'Body, corporeal space, and legitimating chronic illness: women diagnosed with M.E.', *Antipode*, 31, 4, 372–397.
—— (2002) *Women, Body, Illness: space and identity in the everyday lives of women with chronic illness*, Lanham MD: Rowman & Littlefield.
Mother and baby (2006) Available HTTP <http://www.motherandbabymagazine.com/> (accessed 2 October 2006).
Mother Store (2006) Available <http://www.themotherstore.co.nz/> accessed 12 September 2006
Mummy Maternity (2006) Available <http://www.mummymaternity.co.nz/> (Accessed 12 September 2006).
Nash, C. (2000) 'Performativity in practice: some recent work in cultural geography', *Progress in Human Geography*, 24, 4: 653–64.
Nast, H. (1998) 'Unsexy geographies', *Gender, Place and Culture*, 5, 2: 191–207.
—— (2000) 'Mapping the "Unconscious": racism and the oedipal family', *Annals of the Association of American Geographers*, 90, 2: 215–55.
Nast, H. and Pile, S. (eds) (1998) *Places Through the Body*, London: Routledge.
National Childbirth Trust (2006) Available <http://www.nct.org.uk/> (accessed 2 October 2006).
Nelson, L. (1999) 'Bodies (and spaces) do matter: the limits of performativity', *Gender, Place and Culture*, 6, 4: 331–53.
Nemba, K. (1994) 'Induced lactation: a study of 37 non-puerperal mothers', *Journal of Tropical Pediatrics*, 40: 240–242.
Nettleton, S. and Bunton, R. (1995) 'Sociological critiques of health promotion', in S. Nettleton and R. Bunton (eds) *The Sociology of Health Promotion: critical analyses of consumption, lifestyle and risk*, 41–59, London: Routledge.
Neville-Jan. A. (2004) 'Selling your soul to the devil: an autoethnography of pain and pleasure and the quest for a child', *Disability and Society*, 19, 2: 113–27.
New Zealand Herald (16 October 2002) Tapu Misa: wweaty, messy and not all that pretty — that's childbirth, A17.
New Zealand Herald (18 November 2004) 'Mother who breastfeeds pup "didn't want to waste it"', A5.
New Zealand top media — magazines (2005) Available <http://www.business.vu.edu.au/bho2250/Top20Media/TopmediaNZ.htm#Magazines> (accessed 14 October 2004).
Niessen, S. and Brydon, A. (1998) 'Introduction: Adorning the body', in A. Brydon and S. Niessen (eds) *Consuming Fashion: adorning the transnational body*, pp. ix–xvii, Oxford: Berg.
Oakley, A. (1979) *Becoming a Mother*, Oxford: Martin Robinson.
Odent, M. (1986) *Birth Reborn*, New York: Fontana.
Ogonowska-Coates, H. (2004) *Born. Midwives and women celebrate 100 years*, Christchurch, New Zealand College of Midwives.
Oliver, M. (1990) *The Politics of Disablement*, Basingstoke: Macmillan.
Pain, R. Bailey, C. and Mowl, G. (2001) 'Infant feeding in North East England: contested spaces of reproduction', *Area*, 33, 3: 261–72.
Panelli, R. (2004) *Social Geographies. From difference to action*, London: Sage.
Papua New Guinea (2003) Available <http://www.uq.net.au/~zzlhiess/papua_new_guinea.htm> (accessed 3 May 2003).
Park, R.E. and Burgess, E.W. (1925) *The City: suggestions for investigation of human behaviour in the urban environment*. Chicago: University of Chicago Press.

Papps, E. and Olssen, M. (eds) (1997) *Doctoring Childbirth and Regulating Midwifery in New Zealand: a Foucauldian perspective*, Palmerston North, New Zealand: Dunmore Press.

Parr, H. (1999) Bodies and psychiatric medicine: interpreting different geographies of mental health, Butler, R. and Parr, H. (eds) *Mind and Body Spaces: geographies of illness, impairment and disability*, London: Routledge, 181–202.

——— (2002) 'New body-geographies: the embodied spaces of health and medical information on the Internet', *Environment and Planning D: Society and Space*, 20, 1: 73–95.

Philo, C. (1992) The child in the city, *Journal of Rural Studies*, 8, 2: 193–207.

——— (1987) 'The Same and the Other': On geographies, madness and outsiders. Occasional Paper no. 11, Loughborough University of Technology Department of Geography.

Phrase Finder (2006) Available <http://www.phrases.org.uk/meanings/211850.html> (accessed 1 November 2006).

Phua, L. (1998) 'Negotiating power relations in everyday spaces: the politics of the pregnant body in Singapore', unpublished thesis, National University of Singapore.

Phua, L. and B. S. A. Yeoh (2002), 'Nine months: Women's agency and the pregnant body in Singapore', in B. S. A. Yeoh, P. Teo and S. Huang (eds) *Gender Politics in the Asia-Pacific Region*, London: Routledge, pp.1–16.

Pile, S. (1996) *The Body and the City: psychoanalysis, space and subjectivity*, London: Routledge.

Pilnick, A. (2002) *Genetics and Society: an introduction*, Buckingham: Open University Press.

Pinky — mychild.com (2005) Available <http://www.pinky-mychild.com/features/pregnancy/keeping.html> (accessed 20 June 2005).

Pratt, G. with the Philippine Women Centre (1998) 'Inscribing domestic work on Filipina bodies', in H. Nast and S. Pile (eds) *Places Through the Body*, 283–304, London: Routledge.

Pregnancy — info.net (2006) Available <http://www.pregnancy-info.net/pregnancy_and_work.html> (accessed 22 February 2006).

Pregnancy shrinks brain (2006) Available <http://www.personalmd.com/news/a1997011304.shtml> (accessed 29 September 2006).

Prilleltensky, O. (2003) 'A ramp to motherhood: the experiences of mothers with physical disabilities', *Sexuality and Disability*, 21, 1: 21–47.

Pritchard, M. (2005) Can there be such a thing as a 'wrongful birth'? *Disability and & Society*, 20, 1: 81–93.

Probyn, E. (2000) *Carnal Appetites: FoodSexIdentities*, London: New York.

——— (2003) 'The spatial imperative of subjectivity', in K. Anderson, M. Domosh, S. Pile and N. Thrift (eds) *Handbook of Cultural Geography*, 290–99, London: Sage.

Purdy, L. (1996) *Reproducing Persons: issues in feminist bioethics*, New York: Cornell University Press.

Ram, K. and Jolly, M. (eds) (1998) *Maternities and Modernities: colonial and postcolonial experiences in Asia and the Pacific*, Cambridge: University Press.

Randolph, J. (1998) 'That Barbie-Doll Look: A psychoanalysis', in A. Brydon and S. Niessen (eds) *Consuming Fashion: adorning the transnational body*, pp. 177–188, Oxford: Berg.

Reekie, G. (1997) 'History and the bodies of the illegitimately pregnant woman', *Australian Feminist Studies*, 12, 25: 77–89.

Riley, M. (1994) *Māori Healing and Herbal: New Zealand ethnobotanical sourcebook*, Paraparaumu: Viking Sevenseas.

River Ridge East Birth Centre (2006) Available <http://www.riverridgeeastbc.co.nz/> (accessed 4 September 2006).

Rodgers, J (2006) *The Disabled Woman's Guide to Pregnancy and Birth*, New York: Demos Medical Publishing.

Rodgerson, G. and Wilson, E. (eds) (1991) *Pornography and Feminism: the case against censorship*, London: Lawrence and Wishart.

Rose, D. (1993) 'Local childcare strategies in Montréal, Québec: the mediations of state policies, class and ethnicity in the life courses of families with young children', in Katz, C. and Monk, J. (eds) *Full Circles. Geographies of women over the life course*, London: Routledge: 188–207.

Rose, G. (1993) *Feminism and Geography: the limits of geographical knowledge*, Minneapolis: University of Minnesota Press.

——— (2001) *Visual Methodologies*, London: Sage.

Ross, S. (2005) *Birth Right*, Auckland: Doubleday.

Rothman, B.K. (1982) *In Labour: women and power in the birthplace*, London: Junction.

Rubin, G. (1984) 'Thinking sex: notes for a radical theory of the politics of sexuality', in, C. Vance (ed.) *Pleasure and Danger: exploring female sexuality*, London: Pandora, pp. 267–319.

Ruddick, S. (1989) Maternal Thinking: towards a politics of peace, Boston: Beacon.

RYT Hospital — the first male pregnancy (2006) Available <http://www.malepregnancy.com> (accessed 14 April 2006).

Saffron, L. (1998) 'Raising children in an age of diversity — advantages of having a lesbian mother', *Journal of Lesbian Studies*, 2, 4: 35–47.

Schrotenboer, K. and Solomon Weiss, J. (1985) *Dr. Kathryn Schrotenboer's Guide to Pregnancy Over 35*, New York, Ballantine Books.

Sedgwick, E.K. (1990) *Epistemology of the Closet*, Berkeley: University of California Press.

——— (1993) *Tendencies*, Durham, Duke University Press.

Serota, C. and Kozlow Gardner, G. (1998) *Pregnancy Chic: the fashion survival guide*, New York: Villard Books.

Setten, G. (2004) 'The habitus, the rule and the moral landscape', *Cultural Geographies*, 11, 4: 389–415.

Sharp, J.P., Routledge, P., Philo, C. and Paddison, R. (2000) 'Entanglements of Power: Geographies of domination/resistance' in Sharp, J.P., Routledge, P., Philo, C., and Paddison, R. (eds) *Entanglements of Power: geographies of domination/resistance*. London: Routledge, 1–42.

Sharpe, S. (1999) 'Bodily speaking: spaces and experiences of childbirth', in E. Teather (ed) (1999) *Embodied Geographies: spaces, bodies and rites of passage*, 91–103, London: Routledge.

Shaw, R. and Kitzinger, C. (2005) 'Calls to a home birth helpline: empowerment in childbirth', *Social Science & Medicine*, 61, 11, 2374–2383.

Shields, R. (1996) (ed) *Cultures of Internet: virtual spaces, real histories, living bodies*, London: Sage.

——— (2003) *The Virtual*, London: Routledge.

Sibley, David. (1992) Outsiders in society and space. In K. Anderson and F. Gale (eds) *Inventing Places. Melbourne*, Longman Cheshire, pp. 107–122.

——— (1995) *Geographies of Exclusion: society and difference in the West*, London: Routledge.

——— (1981) *Outsiders in Urban Societies*. New York, St. Martins.

Skelton, T. and Valentine, G. (eds) (1998) *Cool Places: geographies of youth cultures*, London and New York: Routledge.

Smith, D.M. (1997) 'Geography and ethics: a moral turn?', *Progress in Human Geography*, 21, 4: 583–90.

Smith, D.S. (2000) *Moral Geographies: ethics in a world of difference*, Edinburgh: Edinburgh University Press.

Smith, J. (2003) The myth of the Yummy Mummy, *Weekend Herald*, January 11–12, pp. G2.

Smith, J. (1974) *Tapu Removal in Māori Religion, Memoirs of the Polynesian Society*, 40, Wellington: Polynesian Society.

Smith, N. (1993) 'Homeless/global: scaling places', in J. Bird, B. Curtis, T. Putnam, G. Robertson, and L. Tickner (eds) *Mapping the Futures: local cultures, global change*, London: Routledge, 87–119.

Snowden, R. F. and H. Deem (1951) *From the Pen of F. Truby King: chapters compiled from the writings and lectures of the late Truby King*, Auckland: Whitcombe and Tombs Limited.

Spain, D. (1992) *Gendered Space*, Chapel Hill: University of North Carolina Press.

Spoonley, P. (1993) *Racism and Ethnicity*, Auckland: Oxford University Press.

Stables, J. and Smith, F. (1999) 'Caught in the Cinderella trap': narratives of disabled parents and young carers, in R. Butler and H. Parr (eds) *Mind and Body Spaces: geographies of illness, impairment and disability*. London: Routledge, 256–268.

Statistics New Zealand (2004) Demographic Trends 'Maori birth rate 2000'. Available <www.stats.govt.nz > (accessed 16 March 2005).

Stearns, C.A. (1999) 'Breastfeeding and the good maternal body', *Gender and Society*, 13, 3: 308–25.

Strollers.com (2006) Available <http://www.strollers.com/> (accessed 2 October 2006).

Stuff — National News (16 October 2002) Porn film maker intends to shoot birth in private hospital. Available <http://www.stuff.co.nz> (accessed 16 February 2005).

Sullivan, M. (1995) 'Regulating the anomalous body in Aotearoa/New Zealand', *New Zealand Journal of Disability Studies*, 1, 9–28.

Sullivan, N. (2003) *A Critical Introduction to Queer Theory*, Circa: Armidale, pp. 43–4.

Sutherland, L. with Brooks, A. (1989) *Pregnant and Chic*, New York: Workman Publishing Company.

Taormino, T. (4 October 2002) Hot Mamas: Pregnant women are hot. The Village Voice. Available <http://www.villagevoice.com/issues/0241/taormino.php> (accessed 10 August 2005).

Taylor, J. (1997) 'The emerging geographies of virtual worlds', *The Geographical Review*, 87, 2: 172–92.

Teather, E. (ed) (1999) *Embodied Geographies: spaces, bodies and rites of passage*, London: Routledge.

The Press (21 October 2002) 'Repulsive sleaze', *Edition* 2, 8.

Thomas, C. (1997) 'The baby and the bath water: disabled women and motherhood in social context', *Sociology of Health & Illness*, 19, 5: 622–43.

Tregear, E. (1926) *The Maori Race*, Wanganui: A. D. Willis.

Tunnah, H. (10 June 2005) 'Child Smacking bill on election agenda'. Available <http://www.nzherald.co.nz> (accessed 24 June 2005).

Two hearts, one soul (2003). Available <http://www.twoheartsonesoul.com/lactation.htm> (accessed 2 May 2003).

Tyler, I. (2001) 'Skin-tight: celebrity, pregnancy and subjectivity', in S. Ahmed and J. Stacey (eds) *Thinking Through the Skin*, London: Routledge, 69–83.

Underhill-Sem, Y. (2000) '(Not) speaking of maternities: pregnancy and childbirth in Wanigela, Oro Province, Papua New Guinea', unpublished thesis, University of Waikato.

Union of Fathers (2006) Available <http://www.uof.org.nz/> (accessed 8 December 2006).

Valentine, G. (1996) '(Re)negotiating the 'heterosexual street': lesbian productions of space', in N. Duncan (ed) *BodySpace: destabilizing geographies of gender and sexuality,* London: Routledge, 146–155.

——— (1997) Angels and devils: moral landscapes of childhood, *Environment and Planning D: society and space*, 14, 5, 581–599.

——— (1999) What it means to be a man: the body, masculinities, disability, Butler, R. and Parr, H. (eds) *Mind and Body Spaces: geographies of illness, impairment and disability,* London: Routledge, 167–180.

——— (2001) *Social Geographies: Space and Society,* Harlow: Prentice Hall.

Valentine, G. and Holloway, S.L. (2001a) 'On-line dangers? Geographies of parents' fears for children's safety in cyberspace', *The Professional Geographer,* 53, 1: 71–83.

——— (2001b) 'Technophobia: parents' and children's fears about information and communication technologies and the transformation of culture', in E. Hutchby and J. Moran-Ellis (eds) *Children, Technology and Culture: the impact of technologies in children's everyday lives*, London: Routledge, 58–78.

——— (2001c) 'A window on the wider world? rural children's use of information communication technologies', *Journal of Rural Studies*, 17, 4: 383–94.

——— (2002) 'Cyberkids? exploring children's identities and social networks in on-line and off-line worlds', *Annals of the Association of American Geographers*, 92, 2: 302–19.

Vanity Fair (August 1991) 'More Demi Moore', cover page.

Vogel, L. (1990) Debating difference: feminism, pregnancy, and the work place, *Feminist Studies*, 16, 9–32.

Waikato Hospital (2005) Available <http://www.waikatodhb.govt.nz/WDHB/default.asp?content=9> (accessed 28 November 2005).

Waikato Times (16 October 2002) 'MP makes right decision on birth', edition 2, 6.

Waikato Times (23 October 2002) 'Children have a right to protection, edition 2, 5.

Waikato Times (31 October 2002) 'Baby named Brianna', edition 2, 3.

Waikato Times (6 November 2002) 'Hamilton survey searching for 30 mothers-to-be', edition 2, pp. 8.

Waikato Times (14 October 2006) 'Kids having kids', edition 2, pp. D1.

Wardell Morrone, W. (adapted for New Zealand working mothers by Merren Parker) (1984) *Pregnant While You Work*, Auckland: Fontana/Collins.

Weeks, J. (2003) *Sexuality*. Second Edition. London: Routledge.

——— (1995) *Invented Moralities: sexual values in an age of uncertainty,* Cambridge: Polity Press.

Weisman, L.K. (1992) *Discrimination by Design: a feminist critique of the man-made environment*. Urbana and Chicago: University of Illinois Press.

Welcome to Waterford Birth Centre (2005) Available <http://www.waterfordbc.co.nz/> (accessed 7 June 2005).

Wetton, K. (2003) 'Parlour patrons suckers for teat treat', *The Daily News*, 18 July 2003.

Whitford, M. (1991a) *Luce Irigaray: philosophy in the feminine*, London: Routledge.

Whitford, M. (ed) (1991b) *The Irigaray Reader: Luce Irigaray*, Oxford: Basil Blackwell.

Wikipedia Kirtsy Marshall (2006) Available <http://en.wikipedia.org/wiki/Kirstie_Marshall> (accessed 5 September 2006).

Wikipedia Male Lactation (2005) Available <http://en.wikipedia.org/wiki/Male_lactation> (accessed 30 May 2005).

Wiles, R. (1994) The impact of pregnancy on fat women's body image: in S. Wilkinson and C. Kitzinger (eds) *Women and Health: feminist perspectives,* London: Taylor and Francis pp. 33–48.

Williams, H.W. (1971) *A Dictionary of the Maori Language*, Wellington, New Zealand: A.R. Shearer, Government Printer.

Women's Health Update Magazine (2001) 'Warrior princess in stunning new role', Vol 6, No. 3, Available <http://www.lucylawless.info/articles/womenshealth/> (accessed 5 September 2006).

Wright, E. (ed) (1992) *Feminism and Psychoanalysis: a critical dictionary*, Oxford: Blackwell.

Wycherley, G. (16 October 2002) 'Pornographer looks for way to beat ban', *New Zealand Herald.*

Young, I.M. (1990) *Throwing Like a Girl and Other Essays in Feminist Philosophy and Social Thought*, Indianapolis: Indiana University Press.

Index